Deep Calls to the Deep

Lance Boyd Fuller

Deep Calls to the Deep

James Boyd Fuller

Published by James Boyd Fuller, 2023.

While every precaution has been taken in the preparation of this book, the publisher assumes no responsibility for errors or omissions, or for damages resulting from the use of the information contained herein.

DEEP CALLS TO THE DEEP

First edition. July 25, 2023.

Copyright © 2023 James Boyd Fuller.

ISBN: 979-8223055518

Written by James Boyd Fuller.

Deep Calls
to the
Deep

*There **is** a deep out there, a kind of cosmic library, what would be in that library? There is a deep to be found. How do we find it?*

We must first be honest and willing to admit we may be wrong of all we ever learned and rely on faith to see that deep meaning, Deep intent. James Boyd Fuller

The Helix Nebula, NGC 7293, lies about 650 light-years away in the constellation of Aquarius. It is one of the closest and most spectacular examples of a planetary nebula. The nebula's striking appearance earned it the nickname the Eye of God

Dedication

This book is in dedication to the lively and wonderful citizens of planet earth our home and our futures. I am forever in debt to the substantial education and learning provided by internet and networks of specific databases in science which hold reality points and truths of our civilization, past and present. I want to thank my wife Barbara, my companion for enduring me and my many times of trials. I thank my family members for being honest and willing to give advice for many subjects. And finally, I thank God and his son Jesus Christ for allowing me this transmission of information in word and spirit of a better future with forgiveness and understanding. I hold no one on this planet in judgment. I have always tried to be open hearted with consideration to important matters which define us together. All chapters contain factual accounts of factual events and verbatim opinions. My own opinions are weighed upon the evidence, the facts, and the calculations in areas of power generation formulas.

I thank everybody on this planet for considering my evidence and future 100+ year plan which rights the ship and repairs the tiller that we steer a proper course again in planetary health, well-being and not being ashamed of who we are.

The Bilderberg council was created in 1954 as a submittal council which is to take control of matters on earth to a new level, but technology has mostly held them from completing their agenda of silent weapons for quiet wars. When you look at the structure of the thinking you see its meant to control all

aspects of human life in the agenda completion. You will see why this matters now and the bottle neck closing all choice corridors.

⌘

Table of Contents

Contents

Introduction

In the face of future catastrophe, humanity seeks answers to deep questions not being asked of our species ability to understand when change is needed to avert natural resource depletion/destruction. Even MIT has fallen on deaf ears.

Most importantly what kind of change is required now?

Political? Environmental? Educational? (Pivot point) Especially, educational? But mostly Social political as of today, i.e., 2022? The political environment has become polluted with greed and happen stance of no real value besides individual class gain in control and power and wealth. Just as the powers in Jesus' day "do as I say," so too have we the same spirit among us today, only now it's different. We now use 1.7 earths because we now have added 2755 billionaires to the resource depletion role. I fully disclose the calculations and implications to this pathway.

The challenge is the *will* to resolve a true problem, *Love of Money* which creates a false sense of security. The delusion of power and money is at its peak today and growing this narrative on all levels of society is dangerous. I have studied why such people with finite lives would pathway a failure outcome and short-term authority experiment. I included the documents of 1954 Bilderburg group establishment of "Silent Weapons for Quiet Wars," which includes details what Robert Fitzgerald Kennedy saw as the CIA's ultimate bondage machine turning all into slaves of the system created to serve the elite and a trimmed down population. JFK was going to expose the details of how in the Dallas TX speech which was never delivered.

With the many scientific debates that define our future course and relevant avenues, we have found, the debate continues as to how to cope with changes happening in natural systems which support and keep us living i.e., air and water etc.

For better part of four decades, this debate has largely gone to no man's land toward blank betterment with Band-Aid corrections like Prince Williams prize that solves nothing except line the power-control venues. The rich come out on top, or do they?

In short, the *pseudo-science* that is trying to be the great hope of mankind has made itself rich with words, lots of words. I'm not talking about the real scientist interested in discovery; I'm talking about the power narrative structure of the technocrats and elite. But now, the scene is catching up with their decision-making. This entire talk going nowhere is costing humanity time, but people are also waking up to reason, logic and purpose. Cosmetic Academics and Billionaires, Oligarchs, Elites own the day, the money and now a failed direction for humanity. Rich live now off of other people's suffering, is there no price to pay? And what is the real cost? I will present my findings later that will open doors to better paths, paths which define purpose and reality.

What could be described as reality as an economic structure has now become a staging run to world destruction through dominance of the ruling classes which allow ignorant type thinking on their part. Why ignorant? Look around, has *growth economics* subsided and found its error? Have the leaders bound themselves to solving social injustice? Not on all accounts, on the contrary these have secured the immanent division of human classes as seen in the truckers of Canada and Trudeau prime minister of Canada and now also in the U.S.

I'm proposing, has mankind dropped the proverbial ball and can't seem to pick it up? Or are these leaders deluded by money as we watch and listen to conference after conference, sideshow after sideshow, you see a glimmer of a direction and consensus then blip, it's gone to endless debate that goes nowhere, or does it? I will bring out why this is being narrated in the fashion of individualist, industrialist wants.

Some voices more than others understand the deep meanings of nature, but do nothing to support change in symbiotic learning which bring solutions thinking because of coercion. This, in my view, is irresponsible and dangerous and those that do it should be fired like UN general secretary Gutierrez, US President Joe Biden for starters for knowing a solution exists and do nothing to provide support in implementation for many nations benefit many peoples benefit. The leaders' decisions are contrary to a proper mitigation pathway.

Where again others simply cannot grasp not taking a lead in an argument that has no winner, simply the focus is lost in individual wants and short sightedness as the likes of Bill Gates or Anthony Fauci .

Scientists are human after all and have feelings too, but to neglect a primary duty to uphold and find truth they fail the system, as the system fails them ! In this book I present "The Root of Why."

Today, the suggesting science is a failure to mankind because science is not being practiced, no, no, no, it's pseudo-science or opinion science has replaced *discovery and truth*. Those who do such things, should lose their post and credentials as leaders immediately on the grounds they fail purposefully to recognize their oaths of conduct as a world citizen. Is this not why we have laws? To protect against such behavior.

I am suggesting those in charge of what cosmetic academics call science, simply lost their focus of what science is supposed to provide in terms of natural conservancy unbiased not to collaborate and get rich off of industry and people's suffering. They started out learning nature as a curiosity like known constants and gravity laws. What happened to change that discovery?

But now, those evident laws are pointers to deeper meanings and science is struggling to find these meanings socially and action based on evidence. The pariah, to hell with pariah when faced with a deeper dilemma survival! Spaceship earth is losing the ability to maintain healthy environments. Science, Politics, and Education are *TOTAL* failures in this respect. Don't think for one minute, I'm talking to the responsible caring. I tell you why all this talk and only action that creates the next disaster and clearly, we have solutions. John Rosebush, a pioneer of waterpower, has solved the EBB problem in water flow kinetic transfer recovery of nearly 90% of the energy produced. Nothing can match this technology today; John will be known for this technology that is now a must. This is the T-Rex downtown New York for technology, and for two years has been challenging the UN and written to President Biden to accept a challenge to protect this technology from the industrialist individualist or billionaire, and put the technology to work for the planet's health and the people as a progression tool into the distant future 100-200 years. Yes, that's very important now that the tiger is in the corner and threaten that tiger, see what happens. Nature is nature, we are nature, and we must learn the relationship quickly.

The need for change has been known for forty + years or longer. In cases of science, in governing bodies which seem to be helpless to make change happen or unwilling. Therefore, we have prepared a one-hundred-year plan, which stimulates the planetary human population by using attribution action venues to show truthful pathways exist for our species and the planet together. We have a way forward.. !!

Our primary motive as a species should be planetary health vis-à-vis breathing, eating etc. Our primary method of choice soils remediation and power generation as well as capital restructure. For instance, by use of microbial life, enhance the soils will not be killed as is the case today artificial fertilizing agents, pesticides, algaecides etc.

When in soils, the microbial is selected specific for that type of soil to reach its potential in mineral base release. This will relieve the burdens on seas, oceans. Lakes, rivers pollution, uptake of Co2, many other benefits will be explained in this book for this pathway.

Nowhere is the destruction more prevalent than around the world in *artificial fertilizers* which kill microbial life in soils by salts. The questions become why? Why do we not see symbiotic transformation as a primary change motivation in educating the Academic world? For over one hundred years, science has gradually understood nature (with a long way to go.) As a species, we are not in line with directives, laws in nature, which allow symbiotic relationships to be known. Why? I will answer why in my further investigation; I devote a whole chapter to this subject. You will be surprised why.

We answered this with actionable mitigation to the worst soils that could be found around mankind's decision making. The answer we found in every case was applied life for life. We applied our methodology to the problem and now after 13 years, the proper formula is evident. Removing the NPK artificial fertilizers, which cause degradation to soils over time by adding salts, which _destroy_ microbial action for plants roots, of which they need to operate, causing death to the soils structure, we become desert makers. We are now desert makers ! On land and Sea. Why do we still use the artificial fertilizers? *Big business of course*, its business before health, its business before natural needs will be met, its *MONEY* before humanity. We are three years before the bottle neck closes in 2025. Common research tells you this intelligence path.

Planetary health degradation is pronounced and growing because of such decision-making in the 21st century. We are all now on notice and responsible for planet earths demise if we continue this course of ignorance and irresponsible behavior. Leaders must wake up, if you think money will save you then you're lost to the pseudo thinking which landed us here today 2022.

I seriously doubt anyone believes what I have just declared for just the soils! The problems run much deeper than the *soils* and *oceans!* The only time anyone really cares to do anything in terms of directional change for humanity is if they are pressured to do so by compelling forces. If the United Nations thinks their quibbles will make lasting change, why then disrespect the people with talk of change when evidence shows <u>*no change*</u>. The structure is not in place to allow for cooperative understanding, we stand as a civilization to end its current existence in the not-too-distant future because of such policies which solve nothing but look good on the outside.

My prediction, the day will arrive where governing bodies such as UN will be laughed at because they themselves would not change and covering their true intentions will be judged for them! Case in point, University of Austin, TX. A new university which is facing realities will leave other universities in the debt of their history they seem to esteem more than change.

We intend to bring positive change, without Cosmetic Academics interference because it is so proven these academics are a failure point in the system of proper teaching guidance governance and knowledge to do so. It would be foolish to continue the outcry to them. This in turn, sends the positive signal of everything is not okay!! If we do not act precisely, we stand to lose much more than people. In fact, everything is out of balance to the point of centuries to repair. <u>*Again Centuries.*</u> Next another possible pathway solution to capitalist system we are using today.

Richard Kiernicki, Time-Equity is a bold statement to replacing failure systems of inequality, "Growth Economics" to know other pathways are available; we will explore Richard's work in chapter 8.

The thinking of proper mitigation techniques and applied technology has not been given its day nor heard by the people, this must end. As I said earlier, a new university of Austin https://uaustin.org is a bold new way of thinking which use logic, seeking truth instead of greed to pave a better human future posterity. The new begins when logic is once again pronounced and active in the thoughts and minds of all citizens around the world. We intend to make this point clear in this book.

If we are to go forward as a species understanding truth, abiding in such pathways and knowing implications of false realities, we are indeed in trouble to meet this new better future. Cosmetic Academics want us to believe, it's okay to destroy soils with artificial fertilizing agents and receive monetary recompense for such decisions and actions. It's okay to use pesticides and herbicides that destroy soils, oceans, seas, waterways, and aquifers. Also, to destroy insects; like bees which without our civilization cannot continue. It's okay to use fossil fuels even with the knowledge of a better system like John Rosebush's Invention and prototype that works.

It's not the science or the scientists doing the work that is running into narrative problems. It's the academics that drive the bus to the school and don't allow the children to get off until the academic dictate the terms. They are being paid very well to lie! I'm using language which points to the problem and this problem will never cease if we as citizens of this planet allow the leaders to seriously undermine correct pathways. We institute correct thinking to establish guidance which considers the importance of a species direction with US, WE, in mind. Such thinking starts with non-aggressive control points, but if the control point is aggressive and undermines truth it's not correct, simple as that.

Why do we fight wars? It's the directive of the current learning to expect to fight wars. We are a more advanced civilization yet do not know the concept of peace which is a responsibility to the future as laid out after WWII and subsequent wars still being fought. The tangible reality of the future must change its base purpose to first recognize error, then act responsibly to correct that error. This book will implicate those who hide behind such values as technology, money, consumerism and inequality which when used as is today keeping the population dumb and non-participatory in government will now be exposed as such. Names and places are left out in most cases, but one's imagination or direct interrogation of the subject roots out the logical sources. We all have choice whether it's for a solution-based project or the telling of a story. That choice is always and should always remain that of the individual without coercion, or mandates of any kind, unless proven by unbiased peer and scientific examination to full extents of time. As they say, the piper pipes a certain sound take warning. Today, that piper's sound is certain, but unclear, and hazardous, take warning and listen to your own voice, heart and mind. Research both or all sides of an argument.

And finally, Jonathan Otto a pioneer of truth. We live in a world where virtually nothing is as it seems. This is especially true when it comes to the COVID pandemic. Most of the information we read, hear or talk about in our daily lives has been filtered and manipulated to make us think and behave in a certain way.

In January 2020, The World Health Organization declared the COVID-19 outbreak an International Public Health Emergency. People were sent into an immediate panicked state. The media reported that this virus, with no cure, was highly infectious and leading a trail of destruction. The fear surrounding this new virus, which no one supposedly had immunity against, caused a total shutdown of multiple economies. The question here is, was this pandemic really due to random events or natural occurrences?

What is the bigger picture here? And what is Big Pharma's endgame?

Let's take a quick moment to look at the process of habituation. When we're repeatedly exposed to certain stimuli, like a car alarm going off in the neighborhood for hours, unless something immediately reminds us of the event, we'll start to tune out and eventually won't hear it anymore. This is the same way the pandemic has been used to "prepare" us to accept unthinkable and unimaginable events. It is clear from our expert speakers and large amounts of research, that the events that have unfolded from thestart of the pandemic have been planned and have been a mechanism of control to carry out a much bigger plan. Jonathan goes on to explain his eBook which I list the address here: http://www.jonathanotto.tv/ The eBook, deep dives into these events, including the development of the rushed, unapproved, experimental vaccines and how boosters are being pushed despite clear evidence that they're not effective or necessary. We'll

look into the real reasons Big Pharma and the government are pushing vaccines and using fear mongering to scare people into accepting the unthinkable. We'll look at how data has been manipulated in multiple ways to send a distorted message to the world and to boost Big Pharma's profits. And we'll talk about the implementation of procedures that, according to research, clearly don't work, including the wearing of masks.

And this huge development March 9, 2022 Dr. John Campbell has been red-pilled[1]

1.　https://www.google.com/

url?q=https%3A%2F%2Femail.mg1.substack.com%2Fc%2FeJxVkk1vozAQhn9NuIH8wVcOPmTTzZaqoUpDorAXZMwQHMAg4zRLfv06zWVXGo-l953RSPOM4AbOg57ZOEzGeaTCzCMwBbepA2NAO9cJdCEr5uMgiEiEnYqhiIiodORU1Bqg57JjzngtOym4kYN6VAd-TAPfaVhd1kRU1A8IRAEVOBacE2RVERCM6_I5lF8rCUoAgy_Q86DA6VhjzDgt6GpBNjYmY61W6kk03nQtJ8NF64mht9ZoX6Xdy9AoV_B-LKHr3IZPbgmgXA2VO8qug2pBN2ZoQS3oC8xvWJDjfCJdm1wGtM0OOL0ndLu_SfFrea82y_H3OgnTbPVne8nJR5ZMSd81ldW2WY7S-wqll5am803yk-28DFK8HuV7dkAfL1ucrm29SnEukzBRP74E3RnRH5ucfo4l8WW981afoXqXp9cD_NRTJdc4yyDP9_HbfiuCVBchr-_t7nbfzStHMoIIQRQThBHxQ494Ig5wHfuIongpghp7vSRniblc-Kg_4_925Gh24_XYUGrN84PXt2qRFfbvr0qauQDFS7skZvQVHPM8iG-

In his latest video, Dr. Campbell looks through the Pfizer documents and concludes, "This has just destroyed trust in authority."

For a man of this stature who trusted the experts, now must assess his mistake or the lies he received from those paid off experts.

A Sub stack reader wrote, "I have watched him for a couple of months. I knew Dr. Campbell would eventually get there and finally two nights ago, he did. He looked stunned. He is obviously a conscientious reporter of health facts. I enjoy how he summarizes and puts out the data. I wondered why he hadn't heard more about all the problems, but he seemed truly convinced in the effective and safe line we've all been sold. I was sad watching it happen though. He basically promoted a very bad shot and for a man with a conscience the repercussions of his advice will haunt him for the rest of his life. I sincerely feel terrible for this poor man."

And another wrote, "That video, in my mind, is one of the landmark moments of the (still unfolding) Era of Covid/Vax Disaster. In it, you see an overall believer in the vexes, and in the basic trustworthiness of the companies and top institutions—though Campbell has been excellent from early on in tearing apart MSM mischaracterization of scientific data and papers—being forced by the same official information his reporting/analysis has always relied on to the conclusion that the bastards lied to us all. He's never at a loss for words, and his politeness to all sides is second to none, but here for the first time, he is stumbling, seems dazed, and yet stoutly continues to walk through to the unavoidable conclusion. Science betrayed him, turned into anti-Science. It led him to falsely reassure hundreds of thousands."

How would you deal with the emotional stress of millions who look to you for advice and truth !?? This is what the book will expose but much deeper realizations are in store.

Chapter 1

The Cosmetic Academic

What is a cosmetic academic? What world do they adhere to? Why do these special academics not use logic or responsible behavior to change? Will they ever change? If they do change, what kind of change will they bring?

These are very short direct questions, which will be explored in this chapter. First, a short history of certain type of academic I will bring to light. The "Cosmetic Academic." This type of academic is distrustful of others and wants to prove a point of their choosing to develop the synoptic future that can be good or bad to humanity. More times than not, it turns out to be bad because mistrust has already proposed the *game plan*. Example of this are many, but the biggest experiment in this current venture is communism or to be precise, Marxism and Capitalism. Yes, it's alive and well even into the 21st century.

What are the Cosmetic Academics doing today?

To answer that, we must look at examples. The University, not to be mistaken for the college, is special in the course of its thinking and promotes its ideology like democrats or republicans represent their constituents' ideals to be elected. The difference with the University is they are not being elected or serve the people; on the contrary, they developed institutional thinking separating what they agree to from what they will not agree to.

What drives the University is how well they appeal to the next generation of thinkers. But what are the next generation taught to think? They are taught to follow the rules and take advantage of your life's situation by learning the *economic system* based on growth. In so doing you should start a business and create a niche which can grow money influence and control with success. The ultimate goal of a university is to convey its heritage as best in class thereby standardizing its ability to exist. Now, you may say that's not all a university does, I'm not delivering every little detail of a university. What I am delivering is the failure point of this type of thinking which is unsustainable in the end because the plan for change is not included. This brings us to who needs to change and why.

The price of not changing has become evident in every aspect of our lives. From family making to how we order importance of things around us are being lost in time and action which produces failure points of humanity, consumerism without recycling 100%. How is this possible? It's how we are taught and the importance of that teaching which is significant to lifelong worth to the whole. For example, thirty years ago the unmistakable warning of overpopulation and planet health was gaining large media attention, yet academics continued as they still do today to propose no significant teaching change which will bring logic forward such as how we become more symbiotic with nature having smaller families or implementing better consumption policies etc.

Don't get me wrong, there are some, the late professor Jansen who looked at climate models and was wowed when he saw the depth of the change worldwide and several others concurred. Dr. Mann and the hockey stick etc.

Early on the IPPC panel 2005 delivered non logical synopsis of the future degradation and denied it was happening!! Today the turnaround of evidence has *forced* these same academics to change their story. Isn't it plausible that these same academics from the 2016 IPPC would hold to not sure yet for climate change reality of which many still refuse to admit failure and call our waste society the product of volcanoes for climate change?

This is dangerous behavior, especially for those who hold high positions and are paid to hold them. The motivation for these academics has become one of live today for tomorrow we die. This has been the thinking for the last one hundred years. This has fostered the implications we are witnessing today and are becoming bound by in time. (One world Gov.) The next evolution in nature will not be as forgiving as an academic purposeful mistake ! How? The recoil effect of nature once it starts you cannot stop the cascade. What cascade? In the arctic we have increased gassing events which are proving to be huge implications. We have a solution to this but will take a cooperative action policy to set in motion. Will that happen?

To foster the better future academics and the oligarch's, billionaire are proposing vaccines to do what? Save us? We are 1.5-2 million or more years as a species and were still here. Leaders on this planet by force they will administer because of their past failures to try and save face. In teaching and logical changes needed to meet future needs will never be considered because they impede elite's agendas. These *cosmetic academics* are the puppet of the money givers.

How do I know this? Bill Gates openly described this in a TED talk in 2005. How 10-15% of the population can be diminished by use of vaccines. At the end of this book, I will list the links available to examine for yourself. I'm not judging Mr. Gates, but I am warning you about intentions here, they point to individual wants and Gates is very good at getting his wants.

Whatever else can be said about this man is probably volumes and really won't further my point as the man is bent on eugenics and everything, he uses including people for pathways his dream on world control and dominance. How many that have succumbed to his will is probably in the millions, and why people still follow his direction is a mystery. His lifelong disasters in India, Africa and Europe through inoculation programs are well known, the man is simply dangerous to the whole thinking in *proper mitigation pathways* !

I postulate to think, if he gets his way more in the future like seeding the atmosphere with Sulfur Dioxide to artificially reflect sunlight away from earth, we are in serious trouble as a species as recorded in my first book "Out of Balance." Let alone natural species that will die with this pathway, he suggests on its fall out. I am one who understands motivation, these cosmetic academics listen to people like Gates because he can flip their bill for them and make life easier less stressful and many cases more glamorous. What is the true cost though? One can not serve the post knowing a certain outcome of disaster is in the future. So, what compels these academics to set aside their real duty and responsibility?

In my first book, "Out of Balance" I described the "comfort zone" in which people in high places of authority will not speak against or about a problem because it would make them a target for ridicule or even their jobs and living, the pariah syndrome. The reality of our world social structure is one of control on every possible level. By the time you were born until you reach the adult age of 18, you are directed to act for the benefit of the economic system, we currently use *capitalism* which is growth oriented and by the way suicidal to our future. The problem with this type of direction is its failure point *snaps*. Those in the way are killed, maimed or worse suffering deaths of starvation, ideological killings and judgment of other cultures. The people in power today are protecting their way of life and look down on the lower classes of people as a slave market. I don't judge these people for their mistakes in the end they will discover their mistake and be judged for them. *The point here is the pathway for each of us as human beings must change to meet the future need not some but all.* If the current academic continues their stride to evade *responsible future need* behavior in teaching correct pathways which bring solutions, then as Einstein had said comes true, *"the definition of insanity is doing the same thing over and over expecting a different outcome or result."*

To be locked into this cycle is in my opinion suicidal because the future need is not being met nor is the society that drives the pathway therefore the error remains, and the masses are locked into that pathway because of the teaching. The solution is simple enough, but the control point presenting that new solution will not allow it. Today we are witnesses to such systems at work delivering false information, false pathways to proper change, and false teaching on the scale of the world today. *The question now, how insane are we?*

I devote a chapter to this question and dive deep into the reality aspects first of why we chose such a direction and second why we keep the charade going!! The academics must realize they are insane or don't care that they are along with all the other leaders. In order to make proper medical treatment on a scale of the world whether they admit they are insane and meet the criteria does not matter, the damage is intensive care at this point. To turn the thinking to that which meets future *needs* of correction is the point. Who is willing to advance this postulate? For the past 20 years I too have been guilty of such neglect and error thinking, what woke me is the consequence and implications to my actions. 18 years ago, I set out to find those reality points which define a better future. I have since found three of seven puzzle pieces that fit the future need. They will be explained in the coming chapters as well as the meaning of change.

My credentials, I went to school but never appealed to the high academic life simply because its flawed, in my view intelligence is knowing what to do with what you *learn*. To apply learning meeting future needs logically which solve problems. If I'm a PhD my responsibility is with the University to raise funds for some research or points of interest, but you must devote lots of time to maintaining that *money flow*.

This to me is a waste of time that I can better serve humanity by using knowledge which satisfies reality not the pocketbook. So, the question becomes do we need money? Do we need to defend a PhD dissertation which can be funded? Or do we use the intelligence to make the points which solve future problems? Must we prove we are smart? How does this solve a problem?

You must always be willing to truly consider evidence that contradicts your beliefs and admit the possibility that you may be wrong. Intelligence isn't knowing everything, It's the ability to challenge everything you know. It's like sincerity, you can be most sincere about a subject to the point of missing if it's wrong. Can we as a civilization afford to be wrong at a time that is most urgent to health matters in nature? No, we cannot afford such behavior in any way or form. The sweeping of the floor takes place to the cleaning of a house or building, we need to sweep out the laggard and dystopia thinking. To the point of those practicing such behavior be fired or removed from the post entrusted to them. There is a large group here. There will not be a second chance, once the cascade begins.

I point out the failure of the academic and educational sectors because it's clear they are a failure to the highest degree possible and that's simply not what humanity is about. Humanity in my opinion is the culmination of past errors making them right admitting wrongs. Have we done this? Why have we gone off the right path onto an insane path which serves to dominate resource consumption on scales never seen in the history of mankind? What drives the focus of consumption to be ever growing and unequal? The answer to these two questions is evidence to a deeper realization which serves to beguile reality. We look at the current system, what dominates? *Capitalist.* Yes, capitalist they dominate because they serve to make life easier or more worry free but what are the costs? But there is a danger also in the capitalist thinking with growth/consumption models.

When we base everything including logic on growth economics what is this pathway concluding? We allow this to grow unchecked. Or changed to meet future requirements which does not happen it's not in the budget. Deplete the system also known as trees, fish commodities, and the health of that environment we are then the loser to the end of such a system. Why are we the loser? Because what is truly important is no longer relevant to the thinking of the society as we see this today. Of course, there are some that do their best to vilify this system, but money serves to make life easier and our species likes that type of living, *to become more dependent on such a system is now where we stand as a civilization* because we rather leave the hard deep questions to the now defunct governments, media, academia, pseudo-science providers of which kicks the deep questions down the silent path. Never to be realized until no more choice becomes the reality. Who harbors the library of right? No one, we are not taught to conserve or balance our thinking which solve the problem of consumption. Very few people are working in the reality forum which expedites a solution base !! Why? Because the intelligence of this planet will not allow it and push all laws to cover their agendas as you will see in coming chapters.

Ask this question, why do governments solve a problem by throwing money at the supposed solution? Because they think the people who make lots of money know the answers. They don't in fact they are the problem and have been for the last 120 years to the modern society we now have today I cannot call modern socially or morally. The governments are now bound or locked into this thinking which is error thinking to begin with on a grand scale, a delusion if you will satisfy the love of money. We now fall into the category of hypocrite irresponsible and deadly.

For example, solar, and wind power generation is intermittent energy, leaves gaps and must be backed up by coal fired power plants or nuclear or geothermal etc. So, the thinking goes like this. We will produce power from solar/wind and put it into batteries for later use. The problem arises when land is used at huge scales to produce a small amount of expensive energy short term energy at the cost of the people by the way. Expensive because over short time the machinery breaks down the land around it dies of frequency and artificial wave energy heat the plants and wildlife displaced. The replacement period is too soon 10-15 years at best in service before major repairs, you're talking very large arrays like 30-40 miles square. Logic calls for a better way but remember the current leader structure has no use for logic and decide to spend the billions yearly at the expense of the local people receiving such power productions services if you can call that a service. Its more like robbery on the scales of depleted health and resource until failure which presents too late scenarios. We have a solution by the way well talk about that later also.

This type of thinking is happening all over the planet because the cosmetic academic says it's a pathway that's viable sustainable. I don't know about you, but I would like to know how that's going to benefit me by raising costs of power production by billions and destroying lands which could be used for farming, trees. We are killing birds, and raising maintenance cost that I must pay for? Is that progress?

Yes, this old thinking is very much being pushed because they only see dollars that they are able to receive. The president of the United States was given written letters telling him, Mr. Biden that there is a better power production pathway which uses no resources after setting up, does not pollute, low cost, low maintenance and he rejected it !! Yes, that's truth so we make that known today.

The general secretary of the UN was given a letter by John Rosebush inventor of the process and was told he; Mr. Gutierrez would alert the member nations via a letter which never did materialize for the benefit of the world and the people for a vote. Why? It's simple logic here Mr. Gutierrez's interests are not for the people or the health of the planet but for the puppet masters he serves the Oligarchs, the billionaire, the destroyers of our world. It's always the ones that seem to be doing the good work on the outside that are the culprits to progress on the truth side. They talk the talk but can't walk the walk or produce action which records good pathways. Feeding people is not the immediate problem. Creating the infrastructure to feed the people is, whether it be farms, education or social attendance. Where is the Pathway? They don't have one !! Not one which meets the needs of the future.

I personally will petition that President Biden owe up to at least why he decided to not support the next generation power production pathway which solves for resource consumption 100%. I personally will petition the UN members to fire general secretary Gutierrez for blatantly saying he would put a letter out to the member nations leaders and did not. It seems to me it has become too easy to neglect the duty and post assigned to leaders of which we rely on for accurate and defining pathways. We can't afford this any longer. It's clear we cannot afford to trust the leaders any longer and steps should be taken to implement lawful orders which remove such impediments to the progress of humanity morally and responsibly. If we must do our jobs, they must do theirs and theirs even more so because their decisions are vital to the prospect of humanities future. Right now, its blind leading the blinded.

Again, I do not judge these leaders' mistakes. I see these are not mistakes but are decisions made that put all of us in harm's way for no real reason other than protecting Oligarchs money flows to special interests that satisfy the lies. If we are to succeed as the leaders are claiming, then the actions would follow which detail such initiatives. We see no such action; we see no logic to their decisions other than money as the only important matter to their agendas of separation and division.

https://www.stopworldcontrol.com/planned is a group of outstanding individuals which are known by their actions, and I support the organization. In time all those who would do harm to humanity will also be brought to trial one way or another. This inference is by their own accord or by the higher authority God.

John Rosebush's Energy challenge to the UN and World. This sets a precedence pathway which solves for many truly sustainable mitigation points. Nearly all the SDGs are met here.

https://www.youtube.com/watch?v=wa2WZcmpzYw
https://www.youtube.com/watch?v=1PMG4ZcNR7s&t=3s
https://www.youtube.com/watch?v=BUbyuKjtHss&t=295s
https://johnrosebush.blogspot.com/2018/03/worldwide-development-corporation.html

Once the truth is brought forward how will these people these cosmetic academics answer to their failures? To propose doing little to nothing these leaders have not understood their positions. What will be the response when the catastrophe of not changing direction cannot be stopped? Now is the time to support the reality pathway our species must now revert to which promotes life on an equal footing not a given by generation after generation of oligarch errors for the gain of wealth, money, power! The following are members of the newly formed CEET or *council of engineers for energy transition*. My question here is transition to what??? I found out what and just about died laughing. Now these are intelligent people but know nothing of common sense or proper mitigation to saving life and planet.

It's so predictable how these persons who have intelligence do not know how to use it properly because they were taught a system, a failure system. When you are gifted with learning and have a drive to find truth and neglect that truth your headed for certain disaster professionally.

Institute for data systems and society MIT Accomplished? We build data-informed models to evaluate the economic and environmental impacts of energy technologies over time and space. Our research aims to accelerate clean energy development by informing decisions made by engineers, policy makers, and private investors for solar projects. Now you see a problem in the translation, intelligence is not all intelligent.

Now, I personally wrote to Ms. Trancik and offered the reality of John Rosebush's invention. Did not receive a peep a thank you or explanation. These kinds of opportunities for real change are being dumped. So, the real reality is ignorance prevails at the highest levels of intelligence and my letter sent via email is as valid then as it is now.

The next subject is Jeffrey Sachs president SDSN sustainable development. Now I'm going to be honest here the man is not thinking in a reality sense or takes the easy money road which weighs the facts and easy life. I venture to guess he was given a good amount of money to spread disinformation about Solar power, Wind power that are destined to fail. Even a child could understand intermittent energy is a waste of time money land resources and learning. The facts are many that the reality of Solar is coming full circle and is failing in every class to secure its future as a solid low-cost power generation venue for humanity in a world sense. Wind is in the same class intermittent failing and hugely costly. I would not trust such systems they cost the public billions in disposal and upkeep and windmills kill birds use lots of land . This is where the road ends for my support of anything the cosmetic academics try to push on the people not for the peoples benefit or the planets, but for the pockets of investors in such failing technologies. When it fails who pays the cost?

When the truth about the cost to the public is revealed on the truthful disclosure the people will reject this technology on the basis its construction pollutes the environment and its waste from end of life is too often not justifying the service life. When you have academics saying it's the only way forward for us is preposterous and a downright lie and I will fight the good fight knowing there are many solutions ahead for our species to hold to.

What I understand is the selling point for solar and wind to the investor has now to blossom, it will blossom but when the blossom finishes it turns to dust and dies. This is what is going to happen to wind and solar it's a pipe dream and holds no water. A Steve Kirsch alumni of MIT could not even speak at a building named in his honor at the MIT campus. Why? MIT does not want to get involved with the Covid19 efficacy debate. Yet they mask and do all the wrong pathways for the students. Is not debate a cornerstone of humanity?

The following are excerpts from Patrick Wood.

Patrick Wood has written two fascinating books on this topic: "Technocracy Rising: The Trojan Horse of Global Transformation"[1] and "Technocracy: The Hard Road to World Order."

As explained by Wood, this new economic system which is not a natural one is not based on common pricing mechanisms such as supply and demand or free commerce. Instead, the economy of technocracy is based on energy resources, which then dictates the types of products being produced, bought, sold and consumed. In essence, energy replaces the concept of money as a commodity.

That's strange enough, but it gets stranger still. Technocracy, which emerged in the 1930s during the height of the Great Depression, the brainchild of which were scientists and engineers, also requires *social engineering* to keep the system working.

If people are allowed to do what they want, consumer demand ultimately drives commerce, but that won't work here. Instead, consumers need to be directed, herded if you will, to consume that which the system needs them to consume, and in order for that to happen, they need to be more or less *brainwashed*. As a result, the technocratic system requires extensive surveillance and artificial intelligence-driven technologies to keep everyone in check.

Technocracy Is Not a Political System

What's more, technocracy seeks to eliminate elected officials and government as a whole. They have no place in this system which, when fully implemented, would run itself more or less automatically, with input at the top by the technological masterminds. There's also no room for nations or nationalism that might influence behavior.

As noted by Wood, Aldous Huxley's dystopian novel "*Brave New World*" offers a compelling glimpse into technocracy. There's no political system. It's all run by engineers and scientists, and the algorithms they create. As noted in the description for "Technocracy: The Hard Road to World Order:"

As a resource-based economic system, Sustainable Development intends to take control of all resources, all production and all consumption on planet earth, leaving all of its inhabitants to be micro-managed by a Scientific Dictatorship.

While the technocratic plan has been underway for decades, things have been rolled out in rapid succession this year. If you've formed the impression that we're all suffering from some sort of "boiled frog" syndrome, you'd probably be right.

Self-evident rights have been stripped from us and people have grown to accept situations that would have been unthinkable a year ago. We've been told to work from home and avoid going anywhere. Our businesses have been shuttered "to protect public health."

We've been told to wear face coverings even while outdoors, while eating and in our own homes. We're now told we'll have to have **vaccine passports**[1] if we want to get on a flight in the future, and world leaders are openly talking about the Great Reset[2] of which will be a disaster at the rate these leaders are not leading.

Now, the central banks were obviously part of this plan too, from the very beginning. The central bank system is crashing as we speak, having reached the end of its functional life as the global debt burden

---

1.   https://articles.mercola.com/sites/articles/archive/2020/11/24/
commonpass.aspx

2.   https://articles.mercola.com/sites/articles/archive/2020/10/28/the-great-reset.aspx

exceeds countries' ability to pay the interest, but the reset they're talking about is not another central bank system.

It will be centralized, yes, but again, the very basis of the global economy will shift away from the commodity of money to the commodity of *energy*. In the interview, Wood explains how the technocratic elite, members of the Trilateral Commission in particular, have influenced and manipulated economic regulations to ensure their success.

Sustainable Development, Agenda 21—It's All Technocracy

As explained by Wood, many of the terms we've heard more and more of in recent years refer to technocracy under a different name. Examples include sustainable development, Agenda 21, the 2030 Agenda, the New Urban Agenda, Green Economy, the Green New Deal and the global warming movement in general.

They all refer to and are part of technocracy and *resource-based* economics. Other terms that are synonymous with technocracy include the Great Reset, the Fourth Industrial Revolution and the slogan Build Back Better. The Paris Climate Agreement is also part and parcel of the technocratic agenda.

The common goal of all these movements and agendas is to capture all the resources of the world the ownership of them for a small global elite group that has the know-how to program the computer systems that will ultimately dictate the lives of everyone. It's really the ultimate form of *totalitarianism*. When one recognizes the source of such thinking you then have the understanding of why these kinds of people think they are better than anyone else at solving or producing the future *Needs* not the populous needs or humanity needs, it's their needs being met ultimately creating the division problem. Moral and Ethical thinking is satisfied in their minds they are doing what is right, when in reality they fail the future ultimate needs of true sustainability as witnessed with the CEET (Council of Engineers for the Energy Transition) group and thousands of other UN contributors which decided they will get the funding and do Solar and Wind.

When they talk about "wealth redistribution," what they're really referring to, Wood notes, is the redistribution of resources from us to them. My previous article, **"The Global Takeover Is Underway**[3]**,"**[1] features a video by the World Economic Forum where they straight up say that by 2030, you will own nothing. *Everything you need you will rent.*

Technocracy 2030—A Glimpse into the Future

A glimpse into this future was also offered in a November 2016 Forbes article written by an unnamed person from the World Economic Forum Leadership Strategy team. It reads, in part:

"Welcome to the year 2030. Welcome to my city—or should I say, "our city." I don't own anything. I don't own a car. I don't own a house. I don't own any appliances or any clothes.

It might seem odd to you, but it makes perfect sense for us in this city. Everything you considered a product, has now become a service. We have access to transportation, accommodation, food and all the things we need in our daily lives ...

In our city we don't pay any rent, because someone else is using our free space whenever we do not need it. My living room is used for business meetings when I am not there.

Occasionally, I will choose to cook for myself. It is easy—the necessary kitchen equipment is delivered at my door within minutes ... Shopping? I can't really remember what that is. For most of us, it has been turned into choosing things to use. Sometimes I find this fun, and sometimes I just want the algorithm to do it for me. It knows my taste better than I do by now ...

The concept of rush hour makes no sense anymore, since the work that we do can be done at any time. I don't really know if

---

3.　　　https://articles.mercola.com/sites/articles/archive/2020/10/23/world-economic-forum-prediction-global-takeover.aspx

I would call it work anymore. It is more like thinking-time, creation-time and development-time ...

My biggest concern is all the people who do not live in our city. Those we lost on the way. Those who decided that it became too much, all this technology. Those who felt obsolete and useless when robots and AI took over big parts of our jobs.

Those who got upset with the political system and turned against it. They live different kind of lives outside of the city. Some have formed little self-supplying communities. Others just stayed in the empty and abandoned houses in small 19th century villages.

Occasionally I get annoyed about the fact that I have no real privacy. Nowhere I can go and not be registered. I know that, somewhere, everything I do, think and dream of is recorded. I just hope that nobody will use it against me. All in all, it is a good life."

However, if you rent everything and have no private property of your own, then who does own all those things? The technocratic elite that owns all the energy resources does. Disturbingly enough, one form of energy resource that modern technocrats apparently intend to harvest, if patents are any indication, is the human body.

Human Body Activity as an Energy Resource

Microsoft's international patent WO/2020/060606 describes a "crypto currency system using body activity data." The international patent was filed June 20, 2019. The U.S. patent office application, 16128518, was filed September 21, 2018. As explained in the abstract:

"Human body activity associated with a task provided to a user may be used in a mining process of a crypto currency system. A server may provide a task to a device of a user which is communicatively coupled to the server. A sensor communicatively coupled to or comprised in the device of the user may sense body activity of the user.

Body activity data may be generated based on the sensed body activity of the user. The crypto currency system communicatively coupled to the device of the user may verify if the body activity data satisfies one or more conditions set by the crypto currency system, and award crypto currency to the user whose body activity data is verified."

The U.S. patent application includes the following flow chart summary of the process. This patent, if implemented, would essentially turn human beings into robots. If you've ever wondered how the average person will make a living in the AI tech-driven cashless world of the future, this may be part of your answer.

People will be brought down to the level of mindless drones, spending their days carrying out tasks automatically handed out by, say a cellphone app, in return for a crypto currency "award." This kind of merging of digital and biological systems is ultimately what "the Fourth Industrial Revolution" is all about. *I'm going to add to Mr. Woods observations here. Ask what are the implications to such structure? Does the Technocracy solve any real problems? I.e. Power production without using resource, Soil contamination, Oceans dying, waterways polluted and thousands more of which I have seen no willingness to change or solve.*

Who Are the Technocrats?

While technocracy used to be an actual private club, the technocrats of today do not necessarily have membership cards, so it can be difficult to correctly identify them all. Key players, however, are the members of the Trilateral Commission, Rockefeller Foundation, other foundations.

You cannot simply join the Trilateral Commission. They select their own members, and it's by invitation only. A list of the members as of 2020 can be found on FredDonaldson.com. Well-known names in the U.S. Trilateral group include David Rockefeller, Henry Kissinger, Michael Bloomberg and Google heavyweights Eric

Schmidt and Susan Molinari, vice president for public policy at Google.

Recognizing the necessity of the media, there's also David Ignatius, a columnist for The Washington Post; David Sanger, chief Washington correspondent for The New York Times; and Gerald Seib, executive editor at The Wall Street Journal.

It's an interesting list of individuals that can be worth reviewing to get an idea of where, how and through whom technocracy is gaining ground and being implemented. Other groups to look at include:

The Club of Rome

The Aspen Institute, which has groomed and mentored executives from around the world about the subtleties of globalization. Many of its board members are also members of the Trilateral Commission

The Atlantic Institute
The World Economic Forum
The Brookings Institute and other think-tanks

Certain individuals can also be identified by their actions. Examples given in the interview include Bill Gates and British Prime Minister Boris Johnson. Once you become familiar with the technocratic agenda, you can start to recognize the players rather easily.

The Plandemics are Part of the Technocratic Takeover Too

The current plandemic is part and parcel of the technocrats' Great Reset that will usher in a whole new world of unimaginable restrictions on freedom. It has already accomplished a massive redistribution of wealth—again, from the middle class, from small business owners, to large multinational companies such as Amazon[4].

---

4.   https://articles.mercola.com/sites/articles/archive/2020/05/23/amazon-empire-the-rise-and-reign-of-jeff-bezos.aspx

Eventually, don't be surprised if you hear talk about providing everyone with a basic income a step toward the 2030 cashless "utopia" where you own nothing and *universal debt forgiveness in return for the forfeiture of all rights to private ownership going forward.*

The lockdowns also had the effect of demolishing local economies around the world an entirely needless manmade situation that is now used as an excuse for why we so desperately need to "reset" the economic system, and while we're at it, we should "build back better."

Lockdowns and school closings have ushered in calls for more online learning, which locks youths into the *digital surveillance matrix* to an even greater extent than before, and as the **COVID-19 vaccine**[5] is being rolled out, it sets the stage for *biometric surveillance, tracking and tracing,* which will eventually be tied in with all your other medical records, digital ID, digital banking and a social credit system.

All of this in turn requires 5G[6], which just so happens to be rolling out in the midst of this plandemic. Again, once you become more familiar with the technocratic agenda, you'll see how all these seemingly random events are not particularly random at all, but weave together, forming a grand net and we are what's for dinner.

The answer is to *first, educate yourself and others.* I urge you to listen to Wood's interview in its entirety, as he goes over a lot that I have not summarized here. Once you can see the plan, the next step is to resist and object to any and all implementations of the technocratic agenda. We can win, for the simple fact that there are more of us than there are of them, but we must be vocal about it and local about it, we need to join forces and present a united front.

---

5.   https://articles.mercola.com/sites/articles/archive/2020/12/08/coronavirus-vaccine-side-effects.aspx

6.   https://articles.mercola.com/sites/articles/archive/2020/04/16/5g-rollout.aspx

I agree with everything Patrick is saying here and the commentary. The warning is factual based and disturbing to understand these technocrats think they are the supreme holders of the future. They hold power and control of many facets of our lives. But we are still many more than they are, and they can be stopped by not using the materials they produce and sell. There will be a list of such companies and owners that shows you a direction to take back control of your life and that of your community. *It must start local !!*

Another example in the health industry quoted from Children's Health Defense and Robert F. Kennedy Jr.

On Nov. 2, the members of ACIP voted 14–0[7] to recommend Pfizer's Emergency Use Authorization (EUA) COVID shot for children 5 –11 years old. Children, children !! They don't get sick with this virus!!

Committee members readily voted "yes" despite many unknowns about long-term safety[8], including a complete lack of data on the risk of heart problems[9] like the ones experienced by some adolescents[10] who received COVID vaccines *Covid Gene Therapy* mRNA using aborted fetal parts.

Neither the disgracefully unscientific[11] vote nor CDC Director Rochelle Walensky's prompt endorsement[12] came as a surprise.

---

7.   https://dailycaller.com/2021/11/02/cdc-acip-panel-vaccine-coronavirus-covid-19-kids-children-5-11/

8.   https://www.fda.gov/media/153409/download

9.   https://childrenshealthdefense.org/defender/fda-moderna-pfizer-covid-vaccine-teens-myocarditis/

10.   https://childrenshealthdefense.org/defender/cdc-397-reports-heart-inflammation-teens-pfizer-vaccine/

11.   https://childrenshealthdefense.org/defender/cdc-advisors-endorse-pfizers-covid-vaccine-kids-5-11/

12.   https://www.cdc.gov/media/releases/2021/s1102-PediatricCOVID-19Vaccine.html

The question comes to mind??? Why should we the people have to watch out for these intelligent beings going off into lala land? Or more precisely cause inequality against laws and people? The forefathers knew of this type treachery would be proposed even in 1770 while fighting the tyrannical British. People, Local jurisdiction, Aware will avert these stupid pathways of inequality mustered by the tyrannical one's greed.

Though billed as "independent[13]," the 14 ACIP members — like the 17 members[14] of FDA's VRBPAC who voted the same way the previous week — *have deep ties to pharma*, with careers[15] [2] that hinge on promoting and rubber-stamping the United States' destructive one-size-fits-all vaccination agenda will become the saying of Jesus Christ's words for the future generation that will see it.

Describing the VRBPAC and ACIP meetings as "a total sham[16]," Children's Health Defense President Mary Holland said, "Sadly, approval from these committees means nothing in terms of safety." *The real agenda is to control all aspects of life as humans on the planet by manipulating the system through MONEY.*

Political scientist Toby Rogers agreed, stating[17] the ACIP meeting "was not a scientific review. It was banal bureaucrats announcing plans for a Blitzkrieg and the bought white coats were cheering them on." The judgment is clear, the higher authority MONEY is now in control. The implications to their decisions will be many and regretful to the end of which the soul is taken and reserved for the fire of hell.

With their vote to give young children the dangerous injections[18], ACIP members signaled that they, too, deserve to be

---

13.	https://www.kvoa.com/townnews/medicine/cdc-advisers-to-vote-on-giving-covid-19-vaccine-to-kids-ages-5-to-11/article_49fe2ffd-a244-5f07-8590-c4e9d0d9fe4c.html

14.	https://childrenshealthdefense.org/defender/fda-pfizer-covid-kids-pharma/

15.	https://www.cdc.gov/vaccines/acip/members/bios.html

16.	https://anthraxvaccine.blogspot.com/2021/11/childrens-health-defense-robert-f.html

17.	https://tobyrogers.substack.com/p/some-thoughts-on-todays-acip-meeting

shunned, along with the powerful institutions with which they are affiliated. The latter include the nation's top universities and leading pediatric hospitals.

Without exception, all the universities at which ACIP members have appointments — Brown[19], Drexel[20], Harvard[21], Michigan State[22], Ohio State[23], Stanford[24], University of Maryland[25], University of Washington[26], Vanderbilt[27] and Wake Forest[28] — have mandated COVID vaccines.

Pediatric hospitals, meanwhile, are playing a frontline role[29] as COVID vaccination sites. Promoting the injection for 5-year-olds, First Lady Jill Biden visited[30] Texas Children's Hospital straight

---

18.   https://childrenshealthdefense.org/defender/vaers-cdc-covid-vaccine-data-injuries-5-year-olds/

19.   https://healthy.brown.edu/vaccinations/faq

20.   https://drexel.edu/coronavirus/health-safety/COVID-19-vaccination/

21.   https://hio.harvard.edu/news/harvards-fall-2021-vaccine-requirements-all-harvard-affiliates

22.   https://msu.edu/together-we-will/covid19-vaccine/

23.   https://safeandhealthy.osu.edu/covid-19-vaccine-requirement

24.   https://healthalerts.stanford.edu/covid-19/2021/10/06/compliance-with-new-federal-vaccination-requirement/

25.   https://umd.edu/4Maryland/health/covid-19-vaccine

26.   https://www.washington.edu/news/2021/06/03/uw-announces-covid-19-vaccine-requirement-for-all-employees/

27.   https://news.vanderbilt.edu/2021/06/09/covid-19-vaccination-requirement-university-issues-additional-guidance/

28.   https://ourwayforward.wfu.edu/faculty-staff/vaccine-faqs/

29.   https://www.cnbc.com/2021/11/10/white-house-says-about-900000-kids-ages-5-to-11-got-a-covid-vaccine-in-first-week.html

30.   https://www.houstonchronicle.com/news/houston-texas/houston/article/First-Lady-makes-Houston-stop-to-applaud-COVID-16620915.php

away, applauding the hospital for the 39,000 pediatric vaccine appointments it had already scheduled[31].

Also worthy of shunning are the 20,000 individual vaccine providers who were pre-positioned to "hit the ground running" and "get shots in little arms[32]."

Within two days of ACIP's and Walensky's verdicts, these providers had administered the jab to thousands[33] of 5- to 11-year-olds, and within the first week, according to the White House, 900,000 children[34] had been injected. The largest experiment on humans ever performed and now the consequences are more deaths less liberty to choose. What is important? Are people important? The results say No !

The indictment is coming. What indictment? Revelations 10:8-11. That little book John eats sweet in the mouth going in because it's God's word, but then the belly is bitter because the wage of sin is death separation forever from life in the truly good form and protected with infallible words love peace. Written words of God to his creation to guide them. When you fly off that truthful road you become ensnared by all the nasty's which were also created not by God but by Satan. Take it however you will, you're a free moral

agent, but remember that little girl with leukemia healed in seconds, how? The point the intent you weigh is truth and garner your

31.   https://www.webmd.com/vaccines/covid-19-vaccine/news/20211104/thousands-kids-covid-vaccine

32.   https://abc7news.com/cdc-approval-vaccine-5-11-santa-clara-county-kids-covid-19-myturn-california-san-jose/11194601/

33.   https://www.webmd.com/vaccines/covid-19-vaccine/news/20211104/thousands-kids-covid-vaccine

34.   https://www.cnbc.com/2021/11/10/white-house-says-about-900000-kids-ages-5-to-11-got-a-covid-vaccine-in-first-week.html

evidence. Assurance from God has always been greater than man or woman. You need belief as a mustard seed of faith that compels you to good reasoning by truthful means rather than what we have become today beguiled !! By money and growth consumer systems which destroy the fabric of life over time no checks no balances. When our civilization realizes the mistakes, it will either be too late or man will have exterminated his own being.

Think about that. Right now, we have the capacity to reverse some of the errors. Once we begin, we can continue mutual changes it's that simple. To reason the future with logic we must again explore the logic which testifies a good pathway, not wanting or greedy misfits which garner division, separation ultimate death of a system on all levels. This is where we can shine the best if we stand against tyrants and vagabonds of time, ever learning and never able to come to the truth. This is the danger for they know nothing of charity only live today for tomorrow we die. This is the false pathway. This is the error of our pathways in education to stewardship principles and preparation of life for life.

Chapter 2

Corporate Infusion to University's

This story is part of Big Ag U, a series by Harvest Public Media[1] and Investigate Midwest[2] on corporate influence at public universities across the Midwest. Harvest Public Media reports on food systems, agriculture and rural issues through a collaborative network of NPR stations throughout the Midwest and Plains. Investigate Midwest is an independent, nonprofit newsroom covering agribusiness, Big Ag and related issues. Read Part 1[3] and Part 2[4] of this series.

Iowa State University had its eye on building its own feed mill.

It would be a working operation, a chance to advance research on how to process corn grown across the state into animal feed. It would also double as a teaching center both for students on campus and for pros in the industry.

But first, the school needed upward of $20 million. University officials could have asked the state legislature for the money. But they knew they would be chasing funds that have decreased over the years and they'd fall in a long line of competitors.

So the university turned to the companies most likely to benefit from what that mill could offer.

That's how the school ended up with the Kent Corporation Feed Mill and Grain Science Complex under construction now. Paid for without a dime from taxpayers.

---

1. http://www.harvestpublicmedia.org/

2. https://investigatemidwest.org/

3. https://investigatemidwest.org/2021/11/15/corporate-money-keeps-university-ag-schools-relevant-and-makes-them-targets-of-donor-criticism/

4. https://investigatemidwest.org/2021/11/16/a-giant-investment-firm-paid-a-university-to-study-one-of-its-biggest-assets-farmland/

Iowa-based Kent Corp[5]., a big player in milling and livestock feed, won naming rights with the leading gift of $8 million. The Iowa Corn Promotion Board[6] and Sukup Manufacturing Co.[7], an Iowa-based company that makes grain storage and handling equipment, also gave large amounts for the project. Iowa State is still fundraising for the project.

"We realized that there was a strong possibility that we could get all of that funding raised through the generosity of private gift donors," said Ray Klein, the director of the Office of Partnerships for the College of Agriculture and Life Sciences. "And that's proven to be pretty much the case."

Agriculture departments across the country find themselves in much the same spot and are adopting similar strategies to replace public money with private donations. That often steers them to donors with ties to agribusiness[8].3

As lawmakers across the country began to cut deeply into tax subsidies for higher education, ag schools have turned increasingly to places where they could get private money — particularly wealthy alumni and companies eager to promote research, training and facilities likely to benefit their industries.

Harvest Public Media and Investigate Midwest found university officials working like sales teams to appeal to private donors and industry when they want new facilities to keep research afloat.

University officials say the shift comes as a nimble reaction to the evaporation of tax dollars that also helps train the next generation of workers in ways the agriculture industry needs. Critics say agribusiness donors effectively commandeer the research of public institutions toward the narrower needs of industry.

---

5. https://kentww.com/

6. https://www.iowacorn.org/about/iowa-corn-promotion-board-directors

7. https://www.sukup.com/

8.   https://childrenshealthdefense.org/defender_category/big-food/

"My fear ... is the corrupting of the sciences, on what research is produced," said Austin Frerick, the deputy director of the Thurman Arnold Project at Yale University.

Private money makes sense for building projects, he said, but he worries that those funders might interfere with research.

"Not only what is produced," Frerick said, "but what is not produced."

'It's about making a profit!!!

Professors like Charles Hurburgh, who teaches agricultural engineering, hope to run the new Iowa State milling complex as a business and as a direct pipeline to the feed and grain industry.

"That's one of the target areas that we want to meet is to be able to help those people get on-boarded into companies," Hurburgh said. "And even existing staff of companies, bringing them up to speed with new technologies."

The practice builds stronger relationships with the farm-related industries where many of the schools' graduates will want to work, Kent Corp. and Iowa State officials said. Mike Gauss, the president of Kent Nutrition Group, said feed mill operators are aging out of the industry.

"And it's important for Kent to really help to make sure that people are being trained, and we're developing that workforce that is coming into those roles," Gauss said. "One of the best ways to retain people is to make sure that people are coming into their roles with their eyes wide open and allowing them to have hands-on real-world experience."

Critics worry an overreliance on industry money steers research away from areas of broader public interest.

"Agribusiness as it stands right now, is not about the small farmer. It's not about rural communities across Iowa," said Emma Schmit, the Iowa organizer for the national advocacy group Food and Water Watch. "It's about Wall Street. It's about making a profit."

Finding money

Kansas State University has a facility similar to the Kent Corporation Feed Mill and Grain Science Complex: The O.H. Kruse Feed Technology Innovation Center[9]. The state of Kansas, the Kansas Bioscience Authority, the university and its agriculture college provided $10 million in funding for the new facility.

But Dirk Maier, who used to work at Kansas State and now serves as the director of the Kent feed mill at Iowa State, said the school "would have never been able to get a facility like that built without private funds."

Maier said projects at universities get funding based on priorities. Though the feed and grain industry is a high priority to the industry, it's a lower priority when compared to other needs of a university. That, Maier said, is where the private money comes in.

Public college funding had fallen more than $3.4 billion nationally between 2008 and 2019, according to an analysis[10] by the Center on Budget and Policy Priorities. The trend continues. Lawmakers in 27 states suggested cuts for fiscal year 2020 or 2021, according to the analysis.

"Even going into the pandemic, funding still remained well below pre-Great Recession levels in most states," Michael Leachman, vice president for state fiscal policy for the center said, " And in many states, the funding cuts were severe."

Oklahoma sits at the bottom of the state funding ladder. Between 2008 and 2019, state lawmakers cut higher education tax dollars by 35.3%[11]. As a result of cuts in Oklahoma, average tuition

---

9.   https://www.grains.k-state.edu/facilities/feed-technology-innovation-center/

10.   https://www.cbpp.org/research/state-budget-and-tax/states-can-choose-better-path-for-higher-education-funding-in-covid

11.   https://okpolicy.org/oklahoma-among-worst-states-for-higher-education-cuts-harming-students-who-already-face-the-greatest-barriers/

at four-year schools grew by $2,153 between that period, according to[12] the Oklahoma Policy Institute.

That's also led to an increased demand for private funding. Most recently, big dollars have come from an Oklahoma State University alumni couple with ties to dairy giant Schreiber Foods. Larry Ferguson is a former CEO of the company.

The Ferguson Family Foundation donated $50 million for a new building[13] on campus for the school of agriculture. The foundation also donated $2 million to help pay for a new dairy barn[14] at the Ferguson Family Dairy Center.

Tom Coon, vice president for agricultural programs at Oklahoma State, said private donations represent a shift in funding strategy for higher education.

"The culture of higher education today is really based on the ability to bring investors to the table who have a private interest, whatever that may be," Coon said. "The way it works is, we've identified what our key needs are, where private funds could really make a difference."

Failing facilities, new money

University agriculture departments across 97 universities have an $11.5 billion dollar backlog in repairs, according to a study[15] released by the Association of Public and Land-Grant Universities. Peter Reeves, who works at Gordian and worked on the study, said investment money is going toward new buildings instead of addressing the backlog of repairs in old ones.

---

12.   https://okpolicy.org/oklahoma-among-worst-states-for-higher-education-cuts-harming-students-who-already-face-the-greatest-barriers/

13.   https://news.okstate.edu/articles/communications/2020/transformational-gift-from-alumni-leads-to-new-name-for-osus-agriculture-college.html

14.   https://news.okstate.edu/articles/agriculture/2020/stotts_dairy-visitor-opning.html

15.   https://www.aplu.org/library/a-national-study-of-capital-infrastructure-at-colleges-and-schools-of-agriculture-an-update/file

"What we ultimately are left with is this situation where, you know, we have aging out infrastructure that has been underinvested in for years," Reeves said. "And the cost of waiting is starting to catch up with these campuses."

Many researchers work in aging labs prone to breakdowns. Some researchers at Oklahoma State work in outdated buildings that haven't been upgraded in decades. Plant and soil sciences professor Brett Carver spends his time breeding wheat inside greenhouses that were built in 1965.

"We're trying to race or perform an Indy 500 in 2021 with a car that would have been manufactured in the (1960s)," Carver said. "We can soup up that car to maybe try to keep up, but we're going to eventually pay the price for that."

Last February, Carver's team paid the price for the outdated greenhouses when the heaters couldn't keep up with the severe winter temperatures. Carver said 30% of his wheat died. That could put him up to a year behind on his research into creating new wheat varieties tailored to Oklahoma growing conditions.

Carver said some aspects of the greenhouses, like dangling ceiling panels and duct-taped doors, can be replaced between seasons of use. But at some point, the repairs can feel like plugging holes in a sinking ship.

"We've done pretty good so far with the 10 fingers," Carver said. "But we're on the tenth finger right now — no fingers left to plug those holes."

Amanda De Oliveira Silva, a small grains extension specialist at Oklahoma State who also spends sometime in the greenhouses, said she needs to quarantine seeds she receives from around the world to follow U.S. Department of Agriculture guidelines to ensure no insects or diseases could tag along. She had to put nets on the windows like a Band-Aid to make sure the greenhouses are properly quarantined.

Randy Raper, the assistant director of the university's agricultural experiment station, said there's a plan to fund and replace the old greenhouses in the next five to seven years. That would take roughly $12-15 million, and he said the school is soliciting private and corporate partners to find the money, he said.

Yet professors like Carver worry about private money interfering with research meant to serve the public.

"It's that obligation that you may have as a researcher to perform research that is of greater interest to the provider of the funds," Carver said. "And that, just by itself, may cut off others who were not part of that funding stream. And as a public provider of genetic resources, I think we need to stay out of that situation."

Carver said the wheat improvement team does get private donations, but they don't prevent his research from being unbiased.

"They are not the kinds of sources ... that would keep us at bay or keep us obligated to a certain faction or certain segment of the industry. We've never had that problem," Carver said. "And I don't anticipate that happening."

Oklahoma State denied Harvest Public Media's record request for private donations made to the college, citing Oklahoma's public records law[16] that exempts universities from disclosing private or future donors.

Coon says 14.7% of the $60 million in research grants from 2019-2021 have been from private sources — 73.7% are from federal and 11.7% were from state agencies.

Of the private grants, some are corporate funders including BASF Corporation, Bayer and Elanco Animal Health, according to supporting documents from Coon. Others are nonprofit organizations such as the Foundation for Food and Agriculture Research, the Noble Research Institute and Oklahoma Wheat Research Foundation.

---

16.   https://www.oscn.net/applications/oscn/DeliverDocument.asp?CiteID=449229

University officials like Coon said that state budget cuts are a tough adjustment, but they are finding a way.

"It's not so different than in a business," Coon said. "You know, we complain about losing some of our state funding, how that's made things stressful but, you know, sometimes when you're in private business, you know, you lose a big customer, or the price of natural gas goes up, and you have to figure it out."

The Questions become many

What we see as models for the future are really a production growth model of business through universities.

This is what I'm pointing out, it's all they are taught and all they know for solution based thinking which in my opinion is seriously erred by its very capitalistic structure. Our civilization is bent on "Economic Growth" which cannot by any means today or the future meet the "*Proper Needs of Nature*" *therefore the destruction of the land the life and the people in due time.*"

What can we do? We can revert to already known venues which satisfy nature needs which satisfy the people over time and generations. The sustainable path is available, but it takes away from the rich, wealthy and that lifestyle of care not for today for tomorrow we die.

My points here are the current leaders simply are not capable of realistic change which meet the future needs because it will cost them and have them sacrifice something they love, money.

But are they really sacrificing all their wealth for a new system of proper mitigation?

They will lose control of these new technologies but their wealth if they so choose will still be theirs. The University seeks money to fund business which teaches a new generation the same error ridden conclusive model. This is not solutions, this is not proper mitigation, this is not survival of the future human race.

The scenario of the elite survive is not possible because growth economics will destroy the planet the money and the people. We see the excess in our societies because the greed is out of control and the elite think they can control such factors as population control methods through fooling people to receive a death dose of vaccine which is designed to eliminate population by means of trickery. When the reality is the factors are not known which raise no questions. The reality is questions are raised and logic of the species inherent ability to sense something amiss will arise. So what is the control point? In an electrical circuit you have control points. Inductance, Capacitance, Resistance. There are many components which can regulate a circuit and knowing these are helpful to controlling the entire structure called electrical force.

There is really no difference between a circuit of electrical force and circuit of control of people with the force of _Money._

This in my opinion has driven the current 20th century into the 21st century without changing to meet the future needs in education and business modeling.

We are simply not prepared for what is coming and to change direction is out of the question with leaders and influential business leaders. So the end result will be catastrophe on all levels of society.

Let me fill in a reality that is happening all around the north and Arctic regions. This story is from November 29' 2021 ;in British Columbia. They should be knee deep in snow by now.

British Columbia extended a state of emergency and fuel rationing until mid-December as the western Canadian province braces for more heavy rain following some of its worst floods on record.

Parts of British Columbia are still struggling with damage from floods and landslides that closed highways and railways two weeks ago, sharply reducing the flow of goods like grain and lumber to Canada's biggest port in Vancouver. While some roads and tracks have reopened sporadically, the damage has largely cut off the country's Pacific Coast from the rest of the country. Some people are still under evacuation orders.

On Monday, the provincial government extended an order that limits non-essential gasoline and diesel purchases to 30 liters (7.9 gallons) in some areas until Dec. 14, from a previous end-date of Dec. 1.

The extension of the order aims to ensure that vehicles providing essential services will not run out of fuel after Trans Mountain, the sole oil pipeline that runs from Alberta to the Canadian Pacific, was shut due to storms. The province has received fuel supplies by sea from the U.S. and via rail from neighboring Alberta, Bruce Ralston, British Columbia's mines and energy minister, told reporters Monday afternoon.

"This has provided a supply of fuel that would usually come from the Trans Mountain Pipeline while the company works toward restarting the line," Ralston said. "We know that a large weather system is expected to hit in the days ahead. What we don't know is what impact that will have on our railways, our roads and the pipeline infrastructure in the province."

Port of Vancouver said that the two main railways that transport cargo to and from the coast have again closed their tracks due to rainfall concerns.

A large portion of British Colombia's coastal areas are on flood watch and warnings with forecasts of up to 200 mm (7.9 inches) of rain in some places through Wednesday.

"It will be problematic because they are coming so close, back-to-back with the run-off and saturated soil," Armel Castellan, a meteorologist for Environment and Climate Change Canada told reporters.

Maintaining oil flow.

Chinese oil and gas giant CNOOC Limited has begun production from the Buzzard Phase II development, offshore the UK sector of the North Sea.

Buzzard Phase II is located approximately 62 miles northeast of Aberdeen with an average water depth of approximately 315 feet. The Buzzard Phase II is an expansion of the already producing Buzzard field.

Buzzard Phase II is expected to reach its peak production of approximately 12,000 barrels of oil equivalent per day in 2022, increasing Buzzard's production to 80,000 barrels in total.

While fully utilizing the existing Buzzard facility, the project has also built a set of underwater production systems. 2 production wells and 2 water injection wells have been brought on stream.

"We are very pleased with the commencement of production at Buzzard Phase II. The constant development of the field will strongly promote the growth of the company's overseas production in the future," Xia Qinglong, President of CNOOC Limited, said.

To remind, phase two of the Buzzard development was sanctioned back in August 2018. It includes a subsea production and water injection manifold located at a new drill center, tied back to the existing Buzzard facility.

It is also worth noting that the massive Pioneering Spirit vessel completed the lift of the Buzzard Phase II topside module onto the Buzzard 'P' platform in late June 2021.

After the topside module was installed, the subsea installation started as well as topside and subsea hook-up and commissioning. All this was delivered before first oil was produced. Another aspect of the Buzzard II installation was the completion of the five-kilometer pipeline bundle for the project in late July 2021 by Subsea 7.

CNOOC Petroleum Europe Limited, a wholly owned subsidiary of CNOOC Limited, is the operator of Buzzard and has a 43.21 percent interest.

The remaining interests are held by Suncor Energy (29.89 percent), Harbor Energy (21.73 percent), and ONE-Dyas (5.16 percent).

Now, I'm not wanting to sound negative here, but to the reality of such behavior shows the desperation for *ENERGY* has become expensive and most importantly follows the same old ideas. We have a solution for worldwide energy needs. The problem is nobody wants to give up control of their assets that pollute the planet on huge scales, deplete planetary resources on wild ventures that 90% of the time cost the companies dearly to operate. This all translates back to the people to pay and by 2024 the cost of doing business in oil will most likely be $150/barrel of oil this translates to pump prices on the consumer side your share 45% increase or $5-6 pump price.

How do they make up for these loses? You and me and Mr. tax or stealing to make ends meet. This next segment is about stealing but by some you would never expect. I was taken back a realization which says, "You will know them by their works."

The widow of the Dean of the Vatican Ambassador Corps uncovers a web of financial and sex abuse cover ups that leads to the top of the Vatican hierarchy.

For almost forty years, Martha Alegría Reichmann and her husband, Ambassador Alejandro Valladares, were among the closest friends of Óscar Andrés Rodríguez Maradiaga, the influential archbishop and cardinal appointed to lead Pope Francis' elite Council of Cardinal Advisers and dubbed the "vice pope" by the Italian media. They were equally close friends with his auxiliary bishop and close companion, Juan José Pineda Fasquelle. They shared their home and their lives with the duo, and Valladares was instrumental in obtaining Maradiaga's appointment to the cardinalate.

In the space of a few short years their illusions about the two prelates would be shattered completely. The Valladares' life savings would be lost in a fraudulent investment scheme promoted to them by the cardinal, who had suddenly begun to act as if he did not know them. After falling afoul of a close male friend of Bishop Pineda, the prelate would refuse to return precious historical documents that they had lent to him and would begin a whispering campaign to end Valladares's ambassadorship.

As Martha Alegría, finally widowed and desperate, searched for answers and sought justice from the Holy See, reporters would uncover evidence that Maradiaga himself had lost over a million dollars in the same financial scam, and that the pair had taken millions more unaccounted-for funds from the University of Honduras and the Honduran government. Even worse, large sums had been spent by Bishop Pineda on close

male friends who lived with him at the bishop's residence, and he was now accused of sexual predation against seminarians.

However, Alegría's worst shock was still to come. After she took her case to the highest prelates in the Vatican, and ultimately to Pope Francis himself, she found that Maradiaga was untouchable, and despite his promises to Alegría, the pope would not, or could not, do anything to correct his lieutenant. Maradiaga was untouchable, enjoying absolute impunity under the protection of Pope Francis. "Sacred Betrayals: A widow raises her voice against the corruption of the Francis papacy" tells Martha Alegría's story of betrayal and her painful realization regarding the true state of the Holy See.

Sacred Betrayals: A widow raises her voice against the corruption of the Francis papacy

"Alegría's testimony regarding Cardinal Maradiaga's involvement of their family in a fraudulent financial scheme, as well as his defense and cover-ups of his auxiliary bishop is deeply shocking. More disturbing is her revelation that Francis continues to protect Maradiaga despite all the misdeeds for which Maradiaga has made himself personally responsible."

- From the foreword by Archbishop Carlo Maria Viganò, the former Apostolic Nuncio to the United States

"This book is dynamite! The compelling testimony of Martha Alegria Reichmann — a widow betrayed by a powerful cardinal she had once considered a close friend — has the power to break down the wall of silence that protects Cardinal Óscar Rodríguez Maradiaga and expose the corruption that reaches to the very top of the Vatican."

- Philip Lawler, editor of Catholic World News, author of *Lost Shepherd: How Pope Francis is Misleading his Flock*

"We have here just one case study of the vast corruption in the Church that flourishes under Pope Francis, and of the Pope's hypocrisy in promoting it. As the author remarks: 'One gets the impression that the devil is gradually taking over the institutions of the Church.'"

- H.J.A. Sire, author of *The Dictator Pope*

Paperback, 175 pages ISBN 978-1-7352671-0-4
Kindle E-Book ISBN 978-1-7352671-1-1
I just wanted to include this little bit of information for the book if you wish to buy one. This woman is a widow and had all that she and her late husband stolen from them by a most trusted friend and clergy !!! This to me speaks volumes as to where we have ended up so lost when it comes to moral behavior.

Pope Francis is simply the Eighth power spoken in the book of revelations. If there were ever a person that fit this description its this pope. I say this openly because it's now a delimiting power against the people. Pope Francis will do anything necessary to limit the damage of obscenities practiced by clergy in secret which has been brought into the light. There are some thousands of cases of which I'm pointing out just one against a widow.

What are the clergy capable of? Every debauchery you can possibly think of is filled up in the past papal reigns up to and including murder of which a whole other book could be written. My point here is you know them by their works rings a loud bell when it comes to academics, clergy, political leaders and media. But why? Why would persons of such positions revert to lies and moral decay. In my opinion it's to protect their status and fear of losing that status of high achievements and intellectual power. Remember most of these Oligarch and rich are those that had it passed down to them upon the demise of the parent.

So, these new control freaks are set on the course to limit the power of the people by controlling the money. It's always been this since the mid 1800's. Now the armament of such control avenues has raised the stakes of these control measures to include a model which will define the future of control.

In all my days of understanding such positions, I still have trouble believing the human species would cause the collapse of the current system to invite the next which is resource growth economics which in its end will fail totally. One cannot calculate nature from a supposed position that uses all available resource until its dead dried up. This is exactly what is taking place as I write these words.

Every business model is not meeting the future need change which satisfies the need of nature and humanities symbiosis. The point here is if the pope and clergy and political leaders all agree that growth economics is the future then buy your grave ticket because there will not be resource left to sustain the model. I estimated and calculated the rate at which this demise scenario would play out and beginning around 2024 these control freaks will have control of the world commerce of which are the banks and the people all of them without a choice as to what they will own buy or sell.

From predatory to predator these controllers will implement the free to kill those opposing them. It's that strong delusion of what future they want not what future is logical to change to and these will succeed because the love of money will abound.

If this sounds a bit scarify, it's not a figment, it's becoming the new reality. Their power against the people power has always been the challenge and now they proceed with impunity.

You do not just shut off the TV and hope it's gone by tomorrow. Here is a fact to ponder. In the year 2000 there were 440 billionaires or that's what the Google said.. ! So now what's happened?

From year 2020 2,050 billionaires and during a Plandemic, there was added 660 billionaires to 2021 if that rate continues, we are toast because it's terms of consumption. If you have just one billionaire how much resource does it take to maintain that billion? Turns out 3 times the resource use consumption. Why? Because we have no waste recapture models on large scale working to recondition that waste and no enforcements of company corporate growth. At the current rate of billionaire consumption, we consume [1.7] earths. You will say not possible we would run out of earth. Yes. And now you see a small point of the argument which will dictate the future without a lot of people in the near future. Yes, the elimination of much of the population is underway today. The engineering of Nano mRNA to cause the demise of those that take it and believe the nonsense our bodies cannot handle a _covid spike protein even if it's man manufactured_. What is concerning is what they the large corporates have in mind next. If we follow the CIA example indoctrination you have total control no choice slavery. Free will is eliminated

We've been a species 1.5-2 million years or more and oh no a covid spike is too much for us to handle? I find such reasoning to be despicable and illogical. But the masses will believe the lie because the leaders promote it as authorities and back it and get rich off of it and control you as a result.

To be more precise the vials are stacked against the weak and frail and useless human for the preliminary phases. Now 2021-2022 they include the children and will weed the weak from these stocks as well.

If I sound perilous, I mean to express the reality first and that the reader can understand the forces at work and how to recognize them. It's not just life at stake it's our very souls the hater of God's creation wants. Satan is not and never will repent; he is bent on destroying all that mankind is because he knows there is no place for him in an acceptable testimony. The following is a small piece of what the Pharmacy industry will take lengths to do. It's totally unbelievable this Pfizer is still in business, but you will also understand the huge money they pull in daily and the *acceptable risk*.

Pfizer Fraud and acceptable risk

Pharmaceutical giant Pfizer, recently known for their development of a COVID-19 vaccine[17], has been caught multiple times engaging in unethical and immoral behavior. This is no secret, yet over the years this fact continues to be brushed under the rug and remain mostly unacknowledged by mainstream media. Since mainstream media has such a large influence over the perception of the masses, it's no wonder why so many people respond to the word "big

---

17. https://www.collective-evolution.com/2021/01/02/50-percent-of-healthcare-workers-in-riverside-county-california-refuse-to-take-covid-vaccine/

pharma" with "conspiracy theory." If one takes a closer look, it's not hard to see why there is actually great cause for concern.

There are many examples to choose from when bringing about awareness to unethical behavior by big pharmaceutical companies, one comes from a paper published in 2010[18] by Robert G. Evans, PhD, Emeritus Professor, Vancouver School of Economics, UBC. The paper, titled "Tough on Crime? Pfizer and the CIHR" is accessible through the National Library of Medicine (PubMed), and it outlines how Pfizer has been a "habitual offender" constantly engaging in illegal and criminal activities. This particular paper points out that from 2002 to 2010, Pfizer has been "assessed $3 billion in criminal convictions, civil penalties and jury awards" and has set records for both criminal fines and total penalties. Keep in mind we are now in 2021.

Evans provides several examples, one coming from September of 2009 when the company settled a number of charges for a total of $2.3 billion (O'Reilly and Capaccio 2009). This particular settlement set a new record for a criminal fine as they pleaded guilty to one count of a felony and misbranding of a pharmaceutical. This means that multiple fraudulent marketing practices were used to promote various drugs. In this case, the criminal charges focused on the "illegal promotion" of several Pfizer brands – Bextra (valdecoxib, a pain medication), Geodon (an atypical antipsychotic), Zyvox (linezolid, an antibiotic) and Lyrica (a seizure medication). These were promoted for uses that were not approved by the FDA and there were also kickbacks to physicians (meaning they got paid for prescribing these drugs).It is reminiscent of Pfizer not reporting a less than 1 percent Absolute Risk Reduction[19] while reporting a 95 percent Relative Risk Reduction for their COVID vaccine.

---

18.   https://pubmed.ncbi.nlm.nih.gov/21532766/

19.   https://thepulse.one/2021/09/03/pfizer-vaccine-offers-less-than-1-absolute-risk-

reduction/

This was by no means Pfizer's first offence. In 2007, Pfizer subsidiary Pharmacia & Upjohn paid $34 million and pleaded guilty to paying kickbacks for formulary placement of its drugs and entered into a Deferred Prosecution Agreement for off label distribution of Genotropin, its brand for the human growth hormone somatropin. In 2004 Pfizer subsidiary Warner–Lambert pleaded guilty and paid more than $430 million to resolve criminal charges and civil liability arising from its fraudulent marketing practices with respect to Neurontin, its brand for the drug gabapentin. Originally developed for the treatment of epilepsy, Neurontin was illegally promoted off-label for the treatment of various forms of neurological pain, and for migraine. Evans goes on to explain how in 2010 Pfizer was ordered to pay $142 million US in damages for fraudulently marketing an anti-seizure drug called gabapentin, which was marketed under the name Neurontin. Pfizer was caught "fraudulently" marketing the drug "and promoted it for unapproved use." It was discovered that the drug was promoted by the drug company as a treatment for pain, migraines and bipolar disorder, even though it wasn't effective in treating these conditions and was actually toxic.

The trials forced the company to release all of its studies on the drug, including the ones it kept hidden. A new analysis of those unpublished trials by the Therapeutics Initiative suggests that gabapentin works for one out of every six or eight people who use it, at best. The review also concluded that one in eight people had an adverse reaction to the drug.

It's quite obvious why the company never wants to go to trial and always ends up paying large sums to settle. Apart from bribing and paying physicians and other medical professionals, the paper points out that they dished out millions of dollars to more than 200 academic medical centers and other research groups for clinical trials. A great quote comes to mind here from Arnold Seymour

Relman (1923-2014), Harvard Professor of medicine and former Editor in Chief of the New England Medical Journal.

The medical profession is being bought by the pharmaceutical industry, not only in terms of the practice of medicine, but also in terms of teaching and research. The academic institutions of this country are allowing themselves to be the paid gents of the pharmaceutical industry. I think it's disgraceful."

Who pays for the pizza? Redefining the relationships between doctors and drug companies. 1: Entanglement[20].

Evans outlines another interesting point which shows why "justice" is never really done and these companies always seem free to engage in this type of criminal behavior. A corporation may treat both criminal and civil penalties as simply business expenses, to be weighed against the revenues earned from illegal behavior. But human beings can be put in jail, and that is a whole other matter. Conceivably, convicting corporate executives of criminal behavior and sentencing them to terms of imprisonment might be a more effective deterrent to the "repeat offender" behavior demonstrated by Pfizer.

These companies are also protected from any harm that comes as a result of their vaccines. For example, the Canadian government announced[21] that it's implementing a pan-Canadian no-fault vaccine injury support program for all Health Canada approved vaccines. This means that pharmaceutical companies cannot be held liable for any vaccine injuries, and compensation from injuries do not come from the company, but from taxpayer money instead. It's like the programs many countries already have in place, in the United States it's called the National Childhood Vaccine Injury Compensation Act. These measures shield and protect pharmaceutical companies

---

20.   *https://www.ncbi.nlm.nih.gov/pmc/articles/PMC1126053/*

21.   https://www.canada.ca/en/public-health/news/2020/12/government-of-canada-announces-pan-canadian-vaccine-injury-support-program.html

and make many of their products, including vaccines, a liability free product. In the US alone nearly $4 billion[22] has been paid out to families of vaccine injured children, and a number of studies are calling into question their safety.

In all of these cases mentioned by Evans, the corporation itself, i.e.., its shareholders – incurred the financial penalties and the executives involved were presumed innocent. Evans states, "In the absence of such personal liability, both criminal and civil penalties appear to be, to Pfizer at least, a business expense worth incurring. You must spend money to make money. Fraud, misconduct, and illegal activity are well-known aspects of pharmaceutical companies' business practices. Unlike other large industries, while business practices may be potentially unethical, but not illegal, those in the pharmaceutical industry routinely and flagrantly engage in illegal activity without facing any deterrent consequences. The Food and Drug Administration (FDA) and the False Claims Act (FCA) deem pharmaceutical companies criminally and civilly liable for engaging in conduct including, but not limited to, misbranding and mislabeling products, promoting products for off-label or non-FDA approved uses, misrepresenting or adulterating data and clinical trial results, and failing to disclose or adequately warn consumers of potential risks and side effects. Violations of these laws and regulations are so widespread and regular, that it is difficult to argue that they are not purposeful. – Anastasia Morairty, Journal of Health and Biomedical Law[23]

Keep in mind that this paper was published in 2010 and only deals with criminal actions of Pfizer from 2002-2010. We are now in

---

22.   https://www.hrsa.gov/sites/default/files/hrsa/vaccine-compensation/data/data-statistics-report.pdf?fbclid=IwAR0P8kUxNcvJB8W_soY6w0T841Kx4ppibrXkle4tZ3ApX96BvmJXGtTUSWo

23.   https://sites.suffolk.edu/jhbl/2020/02/26/avoiding-accountability-the-insulation-of-pharmaceutical-companies-from-criminal-liability/

2021, and the problem has become so widespread that even scientists from within organizations like the Centers for Disease Control (CDC), for example, are blowing the whistle. For example, a few years ago more than a dozen senior scientists from within the agency put out a letter stating the following[24]: We are a group of scientists at CDC that are very concerned about the current state of ethics at our agency. It appears that our mission is being influenced and shaped by outside parties and rogue interests. It seems that our mission and congressional intent for our agency is being circumvented by some of our leaders. What concerns us most, is that it is becoming the norm and not the rare exception. Some senior management officials at CDC are clearly aware and even condone these behaviors. Others see it and turn the other way. Some staff are intimidated and pressed to do things they know are not right. We have representatives from across the agency that witness this unacceptable behavior. It occurs at all levels and in all our respective units. These questionable and unethical practices threaten to undermine our credibility and reputation as a trusted leader in public health. We would like to see high ethical standards and thoughtful, responsible management restored at CDC. A study published in the British Medical Journal in 2016 by researchers at the Nordic Cochrane Center in Copenhagen showed that pharmaceutical companies were not disclosing all information[25] regarding the results of their drug trials. Researchers looked at documents from 70 different double-blind, placebo-controlled trials of selective serotonin reuptake inhibitors (SSRI) and serotonin and norepinephrine reuptake inhibitors (SNRI) and found that the full extent of serious harm in clinical study reports went unreported. These are the reports sent to major health authorities like the U.S. Food and Drug Administration.

---

24.   https://usrtk.org/wp-content/uploads/2016/10/CDC_SPIDER_Letter-1.pdf

25.   https://www.scientificamerican.com/article/the-hidden-harm-of-antidepressants/

Even the pharmaceutical companies have been able to purchase congress. They're the largest lobbying entity in Washington D.C. They have more lobbyists in Washington D.C. than there are congressman and senators combined. They give twice to congress what the next largest lobbying entity is, which is oil and gas... Imagine the power they exercise over both republicans and democrats. They've captured them (our regulatory agencies) and turned them into sock puppets. They've compromised the press... and they destroy the publications that publish real science. – Robert F. Kennedy Jr[26].

Concealing evidence that calls into question various products put out by these companies is quite commonplace. Paxlovid is the first FDA approved oral treatment for COVID under Emergency Use Authorization (EUA.) Pfizer began phase 2/3 trials for the drug with 2246 participants in July 2021. The results, according to Pfizer, show strong evidence that Paxlovid can reduce the chances of severe COVID symptoms by 89 per cent if taken as soon as initial symptoms begin. Ideally within the first five days.

Pharmaceutical company Merck had been developing[27] a similar antiviral drug called Molnupirarvir, however it only appeared to have a 50 per cent reduction in severe COVID cases. The cost of Pfizer's product is about 25 per cent less at around $530 per course versus the $700 per course for the Merck product.

The science behind these antiviral drugs for COVID is based on inhibiting the activity of what is called the 3CL main protease. When this enzyme is blocked it cannot break down the proteins needed to build and replicate the COVID virus in a body. See Pfizer release explaining this further.[28]

---

26.   https://www.youtube.com/watch?v=w9y-OTaJPOQ&feature=youtu.be

27.   https://thepulse.one/2021/10/14/merck-to-sell-covid-pill-for-40-times-what-it-costs-to-make/

28.   https://www.pfizer.com/news/press-release/press-release-detail/pfizers-novel-covid-19-oral-antiviral-treatment-candidate

The idea of blocking the 3CL main protease activity has been on the radar throughout the pandemic. A study[29] published at the beginning of 2021 outlined the best drugs already approved by the FDA that could potentially perform this function. By using computational molecular modeling, a group of scientists were able to analyze over 3000 FDA approved drugs narrowing the search down to 47 candidates.

Here is a part of their conclusion:

"We have identified that boceprevir, micafungin, ombitasvir, paritaprevir, and tipranavir exhibited partial inhibitory effect, whereas ivermectin was able to completely inhibit the SARS-COV-2 3CLpro enzymatic activity in vitro at the tested doses."

From, Identification of 3-chymotrypsin like protease (3CLPro) inhibitors as potential anti-SARS-CoV-2 agents[30]

From, Identification of 3-chymotrypsin like protease (3CLPro) inhibitors as potential anti-SARS-CoV-2 agents[31]

To be clear, Ivermectin is a different molecule than Pfizer's Paxlovid. These drugs are not the same, but it appears they perform the same function. This is important to note because only a new molecular compound can be patented by a pharmaceutical company for up to 20 years.

As soon as this time expires the cost of the drug goes down dramatically because exclusivity is gone, and anyone can manufacture it.

Ivermectin has been used in a number of people for billions of doses and reduces SARS-CoV-2 replication in laboratory studies[32].[5]

---

29.   https://www.nature.com/articles/s42003-020-01577-x

30.   https://www.nature.com/articles/s42003-020-01577-x

31.   https://www.nature.com/articles/s42003-020-01577-x

32.   https://thepulse.one/2021/09/18/ivermectin-reduces-sars-cov-2-replication-in-laboratory-studies-university-of-oxford/

Ivermectin is off patent, therefore the cost to manufacture is about 12 cents per dose.[33]

So as the U.S is set to spend upwards of 5 billion dollars on Pfizer's new drug[34] Paxlovid, it is time to start asking questions.

Does it make sense that an FDA approved drug that has been used by humans for decades with minimal associated serious risks is considered less safe than a brand-new drug created in the last year?

As well under EUA, Pfizer is not liable for any damages their pill causes, giving them a free pass to make lots of money. And what about the fact that Pfizer has been charged billions of dollars for fraud[35], for lying about side effects and data from their clinical trials in the past?

Finally, if there were the resources to create a brand-new drug, why hasn't there been the resources to quickly repurpose already approved drugs that block the 3CL protease? Instead, there have been campaigns to demonize Ivermectin.[36] Governments, media and other authorities like the FDA have actively sabotaged this Nobel prize winning drug.[376]

Experts such as Dr. Pierre Kory, Dr. Joseph Varon and Dr. Peter McCullough have all estimated that 75-80 per cent of COVID deaths could have been prevented with an early treatment protocol that would include ivermectin.

Now Pfizer is saying that their drug, Paxlovid, can reduce severe COVID by 89 per cent if used within the first five days of initial

---

33.   https://www.medrxiv.org/content/10.1101/2021.06.01.21258147v1.full

34.   https://www.reuters.com/business/healthcare-pharmaceuticals/us-govt-buy-10-mln-courses-pfizers-covid-19-pill-529-bln-2021-11-18/

35.   https://thepulse.one/2021/09/17/pfizer-has-been-assessed-billions-in-criminal-convictions/

36.   https://thepulse.one/2021/09/09/breaking-down-cnns-dishonest-coverage-after-joe-rogan-gets-covid/

37.   https://pubmed.ncbi.nlm.nih.gov/34466270/

symptoms. And it just so happens to perform the same anatomical function as Ivermectin.

Truth will come to the surface, and when it does there needs to be people held responsible for withholding and discrediting a drug that could have saved millions of lives.

What will become of moral standards with this type allowance? There will be none, as I said earlier the killing fields will open as depicted in the pale horse of Revelations 6:8 and I looked, and behold a pale horse: and his name that sat on him was Death, and Hell followed with him. And power was given unto *them* over the fourth part of the earth, to kill with sword, and with hunger, and with death, and with the beasts of the earth. *Human Beasts.*

Power was given *them*, who are them? It's those power control freaks corporates that have too much power too much money and a no choice attitude saying they are better than all else on the planet because they have the wealth of the world. There will be no escaping this because the whole world is afflicted with that same inequality of which *money rules is your God*.

I was given a key to share. What is a key? It's the solving of how to open a door or anything that is locked shut up or sealed. What am I opening? Wisdom. The 7 spirits of God are these Love, Peace, Truth, Charity, Compassion, which brings Strength, Power. If we abide by these principles, we cannot lose our heritage or our eternal life, of which I believe is quite present and real. How else do you explain a little girl given up to leukemia (Cancer of the blood) Mayo Clinic the best of the best, frail skinny deathly looking scraggly hair and white all white complexion and being asked if she believed in Jesus Christ? After she is saying yes, immediately healed pink skin, pretty hair, skin filled out and she became hungry asked for something to eat !! More on this later. But to know something exists like that is what hope is to be healed of the sinful nature world that has been built for us to live in for the purpose of choice !!

When God is taken out of the schools, not allowed to be practiced in the open or looked down upon, what could we expect from God in miracles and healings of which are still available today and still happening in places we think cannot harbor God. The lowly, the downtrodden, the peacemakers, those that weep today do so for our benefit, that spirit is not of the dead but of the living !! Then

Revelations 21:7 He that overcomes shall inherit all things; and I will be his God, and he shall be my son.

When God exclaims "He" the reference is to all the living from Adam till today. But what is the promise of the Unbeliever?

Rev 21:8 The abominable, and murderers, and whoremongers, and sorcerers, and idolaters, and ALL liars shall have their part in the lake which burns with fire and brimstone which is the second death. The second death is really the final separation from all love peace truth, can you afford to burn in spirit form forever? How long is forever?

You have a choice to take a chance on man's interpretation by Satan, or interpretation through Gods explanation through Jesus Christ his son which is proven he lives. The truth shall set you free and that saying is true when you lay down the stubbornness of unbelief and practice those 7 spirits of God. How do I know this? How do we have electricity? How do we have knowledge which supports higher understanding? I simply believe God and take him at his WORD because its established and works. He reveals his will through his word when you read it daily things start happening which clarifies life and its meanings. Life for Life. Peace above all with charity. So if in doubt put forward your best effort to at least read for yourself what God has written. If you can't read, there are programs which talk the words. It's that step forward that counts. Then begin you will find truth and it's the world around you being viewed correctly.

There are literally thousands of examples that are available which show what God is doing, what he has done and what he will do. The aspect of trying to find out is a step forward, and Evil is always there to cause you to doubt yourself first then to cause you to think God can't help you. This is not to be a sign of the future you with God, it's a sign your heart is acting on something it wants clarified. In the heart itself there is a chamber called the soul, no one knows what

the function of this chamber is. God said from the heart proceeds the desire, the intent of the body. When we test our heart for discovering good will it not find good? So let us try, let us see what God will do when we follow his word and commands for doing good !!

Chapter 3

Suppression

In the following examples of science and fear we see money is the prime factor of the carnage upon scientist which follow science as it should be practiced. BUT these same scientists are being threatened. One must think a moment and ask, are we on planet earth?

By https://thepulse.one/author/arjunw/

On Nov. 8, 2021, an abstract appeared[1] in the Journal *Circulation* of the American Heart Association (AMA) showing that COVID vaccines "dramatically" increase heart inflammation[2] in people that were in the studies. Plus led to a substantial increase in the risk of heart complications, like myocarditis and heart attacks. Twitter put a note on the post[3] by the AMA, stating that it could be misleading and the study could have errors in it. Cardiologist and NHS consultant Dr. Aseem Malhotra appeared on GBN [4]explaining the findings, and while he was doing so he mentioned another study conducted by a well-known cardiologist, who wished to remain anonymous, that found the same thing. He stated the following,

"A few days ago, after this was published (the abstract), somebody from a very prestigious British institution, cardiologist department, a researcher. A whistleblower contacted me to say that the researchers in this department had found something similar within the coronary arteries linked to the vaccine, inflammation from imaging studies around the coronary arteries. And they had a meeting, and these researchers now have decided that they're not

---

1.　　　https://www.ahajournals.org/doi/10.1161/circ.144.suppl_1.10712

2.　　　https://thepulse.one/2021/11/29/study-suggests-covid-vaccines-dramatically-increase-heart-complication-risks/

3.　　　https://www.rt.com/news/542078-twitter-heart-association-unsafe-vaccines/

4.　　　https://www.youtube.com/watch?v=gJ8t0qQ5R4I

going to publish their findings. Because they are concerned about losing research money from the drug industry. Now this person was very upset about it, I obviously wanted to share this on GB news today."

Dr. Aseem Malhotra

What does this say about the current moment our world is living in? Important information is concealed due to the fact it may threaten one's ability to work, leaving the public uninformed. Pharmaceutical companies not only threaten to stop one's funding if findings go against their business interests, but they also refuse to acknowledge science that calls their products into question !! Now we drop into the pseudo-science realm.

Many of these companies have long had a disregard for ethics and morals. They've even gone so far as to lie about the efficacy and the safety of their products. Robert G. Evans, PhD, Emeritus Professor, Vancouver School of Economics, UBC wrote a paper in 2010[5] [7]titled "Tough on Crime? Pfizer and the CIHR," it is accessible through the National Library of Medicine (PubMed). In it he outlines how Pfizer has been a "habitual offender," constantly engaging in illegal and criminal activities. This particular paper points out that from 2002 to 2010, Pfizer has been "assessed $3 billion in criminal convictions, civil penalties and jury awards" and has set records for both criminal fines and total penalties. Keep in mind we are now in 2021 and these numbers have likely risen. This is concerning, especially given the fact that these companies have big control over academic and medical institutions, as well as medical education.

"The medical profession is being bought by the pharmaceutical industry, not only in terms of the practice of medicine, but also in

---

5.   https://pubmed.ncbi.nlm.nih.gov/21532766/

terms of teaching and research. The academic institutions of this country are allowing themselves to be the paid agents of the pharmaceutical industry. I think it's disgraceful." *I Agree 100% with the assessment, but how to make straight the crooked path?*

What is the premise of such activity? How does a business which kills people continue to be a business? Remember the evidence is greatly against Pfizer and Moderna and Johnson &Johnson and GAVIA etc. The point here is why do the governments allow the killings to take place unabated and unaccounted?

This next excerpt is from Steve Kirsch of Sub stack. If this does not make you angry and mystified at the same time, maybe you're needing to read the proponents of fact and research evidence.

Example 1: mRNA COVID-19 Injections Are Killing Teenagers

The Journal of the College of American Pathologists published a shocking new report on Monday, Feb. 14, 2022 [6]about the cases of two teenage boys who died following mRNA COVID-19 injections.

The report's lead author is Dr. James Gill, the chief medical examiner for the state of Connecticut and the 2021 President of the National Association of Medical Examiners.

---

6.   https://www.google.com/

url?q=https%3A%2F%2Femail.mg1.substack.com%2Fc%2FeJwlkc2OrCAQhZ-m2WlAQHTBYnKTyd3f2Rt-

SiWNQABvx7cf7E5IVYUDdeAroypsMV8yxVLRHZZ6JZABXsVDrZDRWSAvzkpGOB

eDIMhKLAYjNHJlWTPAoZyXKJ3aO6Oqi-E-

zdlEOUO7nC3HXFg90lmRyYzUDjNdCeZcmXVm-

mOqTusgGJDwH_IVAyAv91pTedCvx_Dd1gHZWadCr7yHkDKU0pt4NEUl_065OuOh

VTa6Fgnu-cSnt5JUP-CBdJhR3v27OzIhxHSLX2eNqVzdX1daoeoefdyc6f6o3OxM9-

2CdWErnQvdzysiJ1unAVM8kZmNWPRDD6CFBqUp4atgFvc-

eyOOGB4MHxvpy6lLVeZ5vxdl-VJr2ilt4naje-

82ekvLxxlcvRYISnuwsuYTUP3M5o152SBAbjOzi6qSjGwU49Q-

MmD24XjPaZ7Z3OijZmtjuxVkqQ3r0-

Vi9l9jhKUP&sa=D&sntz=1&usg=AFQjCNGotxpI2MP7nat7oeIT09HGD68xvw

Both boys died in their sleep less than a week after the second dose, and neither had any known health conditions prior to death. In these cases, autopsies of the two teenagers found evidence of myocarditis.

"The myocardial injury seen in these post-vaccine hearts is different from typical myocarditis and has an appearance most closely resembling a catecholamine-mediated stress (toxic) cardiomyopathy. Understanding that these instances are different from typical myocarditis and that cytokine storm has a known feedback loop with catecholamine's may help guide screening and therapy," the report concludes.

This observation is confirmed in another recently published paper by Flavio Cadegiani, "Catecholamine's are the key trigger of mRNA SARS-CoV-2 and mRNA COVID-19 vaccine-induced myocarditis and sudden deaths: a compelling hypothesis supported by epidemiological, anatomopathological, molecular and physiological findings[7]." Dr. Peter McCullough is familiar with Flavio's work and told me he thought it was brilliant.

---

7.   https://www.google.com/
url?q=https%3A%2F%2Femail.mg1.substack.com%2Fc%2FeJxNUk2L5CAQ_TXjrYMxpk
0fPDS9LOxlF2ZgrmK0kkgbFTUT8u-3Mg3Logi--
npVr4yuMMd8yBRLJeej6pFABtiLh1ohk61AVs5K3va9YKIlVlLBjBiJK2rKAKt2XpK0jd4
ZXV0Mp3fPh67nZJGWct4NLRNDN40DHiYE5cCHdkDEsldRvVkHwYCEL8hHDEC
8XGpN5a27v7GfePd9bzIU0NksM7JuAlSE_yuLv67HpLznVD3QxSzR69UFKEpnUHUB
9YRD1ezmGXuKk1rff9_Vx_394_KInxemdLAv7PHn89ePS3tTX9oYzHBxwW4G0HpEo
7N1FZs_vctmLQRlQdcFEWXimsB7F2a1HClizYKeZUsp5orx46EgOQuriz7OyNxjGl3jG
hNm-Iet0YPZvM7ESUYZox0d2hu_UtGwBmAUI-
ixa_tJcEsbn70RK06A03Vum7KNpWrzbJAMyXLXU1q6Do3zKdU3imM7qa5bcBUZBT1
6sLLmDUh97cK3rGqGAFmfxHWV7ZVfxXUQYmCUv3Q79-
J24zdUm2BZGzEqyFJRxqfLxSx_AZxc1uE&sa=D&sntz=1&usg=AFQjCNFylSTzDRBvs
1IH6-0Vw6ZpNRnE_A

So the vaccine killed two kids in Connecticut which has around 1% of the population of the US.

Isn't it odd that we are seeing multiple reports from one small state and complete silence from everywhere else?

Did Connecticut just get unlucky? It appears not.

Let's extrapolate this to the US and estimate around 200 kids in the US have been killed by the vaccine.

Some people may say that that isn't a valid extrapolation.

OK, then try this one for size:

VAERS is underreported by a factor of 41 based on my calculations. This number was recently confirmed by Joel Smalley who analyzed the Massachusetts death data. He got 40.9 (from 1759 actual deaths/43 VAERS reported deaths) which I think is close enough to my estimate.

So 41*64 reported deaths = 2.624 actual excess deaths caused by the COVID vaccines.

Even if you take a very low-ball estimate of the underreporting factor (URF) of 20, you still end up with 1,280 child deaths, a very troubling figure.

So our simplistic 200 estimate just from linearly extrapolating the known myocarditis deaths in CT was an underestimate (as we expected).

Let's be clear. Dr. Paul Offit has said that 1 death per million vaccinated is unacceptable for a vaccine[8]. Here, we have more than

---

8.   https://www.google.com/
url?q=https%3A%2F%2Femail.mg1.substack.com%2Fc%2FeJxVkN2OhCAMhZ9muNMg
IOgFF5ts5jUMQlUyCIafNb794szNbkJp0tP2pJ9WGdYQL3mElNH9Tfk6QHo4k4OcIa
KSIE7WSNb1vSCiQ0ZiQbSYkU3TEgF2ZZ1ER5md1Srb4O_ung20Z2iTZumpYCOnvA
M1YjwzrAaqKRkFF5jDx1QVY8FrkPAD8QoekJNbzkd60K8HedaXcpVeNia9tanMKSv9a
nXYq3TUUFqHaKxfmxwaE5tDFdeEZbG5OaFJWyjOPOiz7vs-
kZUEE4IpHrqRcSxa0gLMYgY1065fBDO4ddFpsQf_YHhfu3-WKMpTLcdGaRXX-
_x3tRKYat6Lt_mawKvZgZE5FkD5w_eNalrBQ6zczaSy7Djjgg9CDASzD4ub9TiysRJE1da

1 *child* death per million vaccinated which is even worse than just 1 death.

Is anyone publicly calling for a halt to these vaccines for kids? No way. Not a chance. They are nearly all pushing for giving it to even younger kids.

Is anyone calling for amending the label on the vaccine to note this new information? Are you kidding? No way. AFAIK, the label hasn't been amended in over a year.

Example #2: The CDC has been withholding unfavorable vaccine data because it might be misinterpreted

On February 20, 2022, <u>*The New York Times* confirmed that the CDC was withholding data</u>[9] that is unfavorable to the safe and effective vaccine narrative.

Kristen Nordlund, a spokeswoman for the C.D.C., said the agency has been slow to release the different streams of data "because basically, at the end of the day, it's not yet ready for prime time." She said the agency's "priority when gathering any data is to ensure that it's accurate and actionable."

Another reason is fear that the information might be misinterpreted, Ms. Nordlund said.

EOuXln_t_AQuZk3I&sa=D&sntz=1&usg=AFQjCNHvBb1xU7V9XoxGrtMCy3SJCvE
B5g

9. https://www.google.com/
url?q=https%3A%2F%2Femail.mg1.substack.com%2Fc%2FeJxVkEGOhCAQRU_T7DQI
KLpgMcmkr2FKKJW0ooFyjLcf7N7MJBQk9av4-c8C4bTFy-
xbInZfPV07moBnWpAIIzsSxt47o6q61kJXzBmuhdUD86kfI-
IKfjFsP4bFWyC_hXu6Vq2sFZuNbhqobD0qUFDbanRSCT1yjdCMwFv9MYXDeQwW
Df5gvLaAbDEz0Z4e8ushnvkkytLLx2TnMh1DIrCv0m5rlvZc1tkC3OopFZ6K09M84-
IKBwTFGPOYfOa_vk_mjeBCcMnbqlMN16UoEQc9IAyyqketHC-XuFi9buGh-DpV_-
xYNCeM-yxlFqc7-rub0_f5XY_g6eoxwLCgMxQPZPRh-
8bUTxgwZuauBzJVoxrdtFq3gqsPh5tz16ku02PZ1m15K5g_2X8BqHeSxQ&sa=D&sntz=1
&usg=AFQjCNGEakFvuHlP3Jj_E4MfuPePqRPBKw

Watch this CBS news video at 7:39[10] where the New York Times reporter says clearly: "One of the big problems I've been hearing about here is that the CDC didn't want to get some of these data out because they were worried it would be misinterpreted. And all the experts I spoke to said that this is a terrible reason because when you hide data or when you keep data from being released it actually breeds more mistrust."

Where is the outrage?

Apparently only from misinformation spreaders like Dr. Robert Malone, Dr. Peter McCullough[11] and Dr. Paul Alexander.

---

10.   https://www.google.com/
url?q=https%3A%2F%2Femail.mg1.substack.com%2Fc%2FeJwlkcuOhSAMhp_msNOAo
OiCxWzmNQzQqmQQDOA4vv3gOUnTJn-v-
Wp1wTWmWx0xF_K4udwHqoBX9lgKJnJmTLMDJVjfy04yAorKzkpDXJ6XhLhr5xU5T
uOd1cXF8FT3YuS9IJuSC4CGxYwjNR2whY-
WIfZ8GnCgIPRnqT7BYbCo8BfTHQMSr7ZSjvziX6_uu9p1Xa01-
TmstXGvyq8DjDVasM0ZAFOzuITNElOjvccVwd_N5cq2RQ8urI2NtaVhUwO66PdU_kc
c6mjXUU5HNomByrZrEY00qA1n_SIF0NYnb-Uew0vQfWVtPk0u2v48d5CkLr0cG-
c1uT4o3mqlMde4n8GVe8agjUdQJZ1Iyof1G9u8YsBUfwCzLooNYpDDKOXYUfHh8nC
fJjFVmqSuhVi7gsqlYvpxKdvtH9xnmMU&sa=D&sntz=1&usg=AFQjCNHJ2XHkhj5IkI
VldVbXGQRIhS3QlA

11.   https://www.google.com/
url?q=https%3A%2F%2Femail.mg1.substack.com%2Fc%2FeJwlkNuOhCAMhp9muBsDiC
IXXOzNvobhUJUMgim4E_fpF3eSpm16-tPPmQprxksfuVRyu7leB-
gE7xKhVkByFsA5eC3YMEguGfGaSu6kJaHMCwLsJkRNjtPG4EwNOd3Tg5j6QZBNO0
apWKSzg5JMOSWEV5ZLAAP9aCb4iJrTB0gONPwAXjkBiXqr9SiP_uvBv5u5UMMvpF
vwQCilc3lvZYtgXiGtLS0Vc1qf74y-PBfM-
9PjczexXWtdEjSnnNOeTkyJkcqOdwBWWjC2Z8MihaddxOjkntND0H1lXTltqca9bi2C-
m2WY-v71lzvn_-r7e25xf1MoV4zJGMjeF3xBFI_UP_5zCskwAbbz6ZqNopRjpOUE6fiA-
AGrJRQDRtpsj63raRLbTxeAYvb_gBBd5Ef&sa=D&sntz=1&usg=AFQjCNH-
2tJqE0uHQUCN8qjC6U0SLpqaAA

I can't recall seeing Robert so upset (you can tell he's upset because he repeats himself over and over at how upset he is). Watch this video[12] which has been viewed fewer than 100,000 times:

Example #3: The silencing of Dr. Peter Schirmacher

I've written earlier about the unethical silencing of one of the world's top pathologists, Dr. Peter Schirmacher[13].

Nobody in the scientific community spoke out how Schirmacher's family was threatened which caused him to be silenced about his stunning results.

---

12.   https://www.google.com/url?q=https%3A%2F%2Femail.mg1.substack.com%2Fc%2FeJwlkcmOhCAQhp-muWFQUfDAYS7zGoalUDIshqU7_faD3QkBastf9ZWWFY6U3-JKpaL72uv7AhHhVTzUChm1Anl3RtBxWdjERmQEYZNmCrmy2wwQpPMCXU15p2V1Kd7ZC-XzQtEpRrMRa0c-W8uV1XK1RpOFa7vxbkv1FZXNOIgaBDwhv1ME5MVZ61Ue889j-u0nt6A8DDqFbjxfxxEWbPKAc1KQKw7S9yoso_l4L-it46B18z6148SyfwtgbTROFhfd1aqzTmObu_Zw1uCRExOZJjITPm50JWyYBgDFFEg1j4t11JDBZ69ZSPFBSTjGoTRVqtR_d18oi5e01znPPXjcVD7eDmbvb2jR1fcOUfYxjKi5Aapf7B-C-wERcl-H2WUV40pXtnLG-EToF9G9gm2jWweLuqxJvSqKUjuxP5eLPv8BO_iebw&sa=D&sntz=1&usg=AFQjCNGh_bJV5jglzNTYMComO68gcsISpw

13.   https://www.google.com/url?q=https%3A%2F%2Femail.mg1.substack.com%2Fc%2FeJxVkMuOhCAQRb9GdhpEFF2wmGTSv2F4VCtpBQLldPz7wXYzk_BI6lLcuscohCWkU8aQkVzHjGcE6eGdN0CERI4MaXZW8rbvBRMtsZIKZoQmLs_PBLArt0kSD705o9AFf73u-dj1nKxSdb0Qw9BP2piJaTuNahTWcka1Bj72t6k6rANvQMIPpDN4IJtcEWOuuq-KPcrKWKSXS9msTT50RmVejQl7kWLZytchgq_voWsM9e6wjgmys-Cx3qruUT77fhMnGWWMdnRsJz5Q0bAGQAsNSndt_xTc0mZLmxF78BWn-9L-8yNJvtUzrl1XxOXK_qmW-HO598M7PGfwSm9gJaYDCN5wP5zmBTykAt3OCmU78EEMoxAjo_wGcYGeJj4VfKTY2lC6vPwT_hcxoJMe&sa=D&sntz=1&usg=AFQjCNGYwBLOe1STTys58alHudyzpcXN7w

Let's be clear: the lack of outrage from the scientific community is a tacit endorsement that physical threats are acceptable for silencing scientists who have evidence that is counter-narrative.

The CDC responds by doubling down on the deception

The CDC just published a study in *The Lancet Infectious Disease*[14] saying that there is no link between the vaccines and death.

I'm serious.

Check out the BBC story that just came out, "Covid vaccines not linked to deaths, major US study finds[15]." The article goes on saying, "A major study of vaccine side-effects in the US found no link

---

14.   *https://www.google.com/
url?q=https%3A%2F%2Femail.mg1.substack.com%2Fc%2FeJwlkc1u5SAMhZ8m7BrxT7Jg0U
2l7kaaB4iAODe0hETgTOa-fUmvhMzRsS1bn4NDeOzlaY-
9IrnDhM8DbIarJkCEQs4KZYqzlUwpww0js6WGB-
NJrNNSADYXkyXH6VMMDuOe72olB6EkWe3CBj1qL5lky6CEUjAIrYNjmo8CvHsNdeccI
Qew8A_Kc89Akl0Rj9qJ945_tHddV48rJNeqsA_71ryv_SzZpdpk82NemnAFY0jQ1J_Pz79M
GvEm6Dh2fOC84yOlVMm3oaWXMyWE_0ii5ZRzKujARqmp6XkP4I0H5wVTi5Ez7VNJ
wWx77iTdHqyvp6_owve9Byn2csuxCtGSjxvGr9t4TO3fzhzxOUF2PsFssZxA8EX7F9z0gAyl
XWGeHFqmpTZ6MGbgVL7I3OTHUY6NJ2lj5711ZVuxgfqOpYb1ByPvlIA&sa=D&sntz=
1&usg=AFQjCNHvCMgm-9b5ypMxo7a3pMz0K67Zgg*

15.   https://www.google.com/
url?q=https%3A%2F%2Femail.mg1.substack.com%2Fc%2FeJwlkE2KwzAMhU9TbwzBf4
mThRdDYS4wzDo4ttJ4mtjBVlpy-
3FbEBI8STze5yzCLeXT7KkgebURzx1MhGdZAREyOQrkMXijeNtqoTnxhmnh9ERCG
ecMsNmwGrIf0xqcxZDi67pVvWwVWYzkUg5tJyfXzTPAbLloue67tnPgmZw_pvbwAaIDA
w_IZ4pAVrMg7hf5dRHfta7pETx9WOdChEJjQrqGeAdPMVEPFpdyEVe62b-
U6e8PLXj4k84h-
kKCEUwIJlnPB9Ux3YgGYNIT2EnydtbKs2bNq9NbihfFthtvyjEVtO7euLSRbJ523hcp6_L
2SvtWa-
Cxzu2IAc8Rop1W8AbzAQQ_ON9kxhtEyBWzHy0a3qlOd73WvWDqE_2FdhjUUIGRa
utT_YqmYCVxD7m45R_yAY41&sa=D&sntz=1&usg=AFQjCNEZU1FVz-
KqJ9zSjk_ccnmNqpdshQ

between two Covid jabs and the number of deaths recorded after vaccination."

No link?!?!

OK, then how do they explain:

The huge increase in deaths reported by multiple insurance companies[16]

Up to a 93% incidence of telltale blood clots[17] noticed by an embalmer.

---

16. https://www.google.com/

url?q=https%3A%2F%2Femail.mg1.substack.com%2Fc%2FeJxVkMuKhDAQRb_G7JSYh9FFFgND_4bkUWromEgSp_HvJ3ZvZqAeULeqLhyjCqwxXfKIuaC7zOU6QAZ4ZQ-lQEJnhjQ7K1nPuSCiR1ZiQYzQyOV5SQC7cl6i49TeGVVcDPc2ZyPlDG2STlypYdEUC44tM5xSWwcDYG0sJ-PHVJ3WQTAg4QfSFQMgL7dSjtzQr4Y8auRSpadL2WxdPnUuyjw7E_cqHTV1VMm2O

-

waUhuX1qu0QrtC2lVoXchnUvV9Qx_14_cLOUkwIZjisZ_YgEVHOgAtNChNe74IZnHnkzdij6FheF_7f6YoyZdajo3SKq43gPe0Mphr38_gyjVDUNqDlSWdgMqH8BvWvEKAVMnbWRXZD2wQwyjESDD70LhpTxObKkNUbW2sV0H-IfALcMaVLQ&sa=D&sntz=1&usg=AFQjCNEzB9V8So0WVIPOfnSJXnj1Muls2w

17. https://www.google.com/

url?q=https%3A%2F%2Femail.mg1.substack.com%2Fc%2FeJxVkMuOhCAQRb-m2Wl4KbpgMcmkf8PwKJU0ggHsHv9-sHszk_BI6lLcuseoAktMp9xjLug6pnLuIAO8sodSIKEjQ5qclZx0naCCICuxoEZo5PI0J4BNOS_RfmjvjCouhut1xwfWcbRKC50SigJmIyhtOBFs4BwbomZGZ9N9TNVhHQQDEp6QzlygAebmWsucb-7rRe125VOnhUjZrmw-dizKP1sStSnvd8GP8kd0TGti08hukJtUG5XMzsibON3avf32_kJMUU4oZHsjIeyxa2gJooetwjHSz4Ba3PnkjthhuHG8L-WeHknypeV8Zq-JyRX9Xa_qp3tsRXDknCEp7sLKkA1D5sH1jmhYIkCpzO6kiSc970Q9CDBTzD4eL8zjysdJD1dbG2hXkn-y_2TaS2w&sa=D&sntz=1&usg=AFQjCNESgtUxJuoc3Fp6Djfjjj6_8aw83Q

The deaths in Connecticut (noted above) which were determined to be caused by the vaccine. These kids died in their sleep. If it wasn't the vaccine, what caused these deaths?

The highly unusual causes of deaths in the kids the CDC analyzed[18]. Those kids did not die from normal causes. But the CDC never mentioned that in their analysis that the causes of death didn't line up statistically. They just said nothing, nothing! (reminding me of the famous line uttered by Sgt. Schultz of Hogan's Heroes fame).

---

18. https://www.google.com/
url?q=https%3A%2F%2Femail.mg1.substack.com%2Fc%2FeJxVkM1uhSAQhZ9Gdhp-
RRcsmjT3NQzCXCVFMIC1vn3x3k2bAJPMYebkfEYXWGK61B5zQfczlWsHFeDMHk
qBhI4MaXJWcSKEpJIgq7CkRs7I5emZADbtvEL7MXtndHEx3L8FH5jgaFU9HYXWloj
eMGmJEZwOhmmgRvR6FPRtqg_rIBhQ8A3pigGQV2spe27YR0Mf9eRSpS-
Xslm7fMy5aPPVmbhVaa_3XK_W6FBaHe7xFn52r11o13i2ZYUMDXvUZZ8ncopiSjHD
Axl5j2VHO4BZzqBnRsRTcos7n7yRWwwNx9tC_vmhpE793FfGqrjc2V_dGn-
qdTuCK9cEQc8erCrpAFTecF-
cpgUCpArdTroo0vNe9oOUA8X8DeIGPY58rPhQtbWxTgX1J_wvD2qTEw&sa=D&sntz
=1&usg=AFQjCNHf9MdKkNi3jbvfcpuIIyFm6hm-Nw

Dr. Peter Schirmacher's study[19] which found the vaccine caused the death in at least 30% of the cases examined (deaths within 2 weeks after being vaccinated).

Dr. Sucharit Bhakdi's study which found that the vaccine caused the death in over 90% of the cases examined

If it wasn't the vaccine, what killed Jacob Clynick[20]?

If it wasn't the vaccine, what killed Ernest Ramierez Jr?

---

19. https://www.google.com/
url?q=https%3A%2F%2Femail.mg1.substack.com%2Fc%2FeJxVkMuOhCAQRb9GdhpEF
F2wmGTSv2F4VCtpBQLldPz7wXYzk_BI6lLcuscohCWkU8aQkVzHjGcE6eGdN0CERI
4MaXZW8rbvBRMtsZIKZoQmLs_PBLArt0kSD705o9AFf73u-
dj1nKxSdb0Qw9BP2piJaTuNahTWcka1Bj72t6k6rANvQMIPpDN4IJtcEWOuuq-
KPcrKWKSXS9msTT50RmVejQl7kWLZytchgq_voWsM9e6wjgmys-
Cx3qruUT77fhMnGWWMdnRsJz5Q0bAGQAsNSndt_xTc0mZLmxF78BWn-9L-
8yNJvtUzrl1XxOXK_qmW-
HO598M7PGfwSm9gJaYDCN5wP5zmBTykAt3OCmU78EEMoxAjo_wGcYGeJj4VfKT
Y2lC6vPwT_hcxoJMe&sa=D&sntz=1&usg=AFQjCNGYwBLOe1STTys58alHudyzpcX
N7w

20. https://www.google.com/
url?q=https%3A%2F%2Femail.mg1.substack.com%2Fc%2FeJwlkU2OhCAQhU_T7DAK
KLpgMZu5huGnuiWDYKDU9O0HuhNCFQ-KevnKaoRXym91pIKkbSu-
D1AR7hIAETI5C-
TVOyWGcZRMDsSpXjlrDfFlfWaAXfugyHGa4K1Gn2J7PYqZj4JsCuwk-
CLtJI1YpoUbY-xclUHaUTi-
fJvq03mIFhRckN8pAglqQzzKg_882G9d93131hTBOpv2em4Oa9hAB9xqYlNOUV8-
n02OicLlXfuS2lQzemlrfQQawFFMFAFioQ40bvTIyQDNtbUOrZh4xXrGet7PwyKmXn
asAzDSgDZ8GJ9SuL4LOVi5p_gQ_f4aunKagtr-
NXckq1s_j43zevlqeD5qJbTWuJ_R43uFqE31ojCfQPDL_4NyfUGEXOfiVo1qmMQkp1
nKmfXiy6rNYlnEUgmT2talWhVVwer_z-
dit38uw59T&sa=D&sntz=1&usg=AFQjCNHNLPwdRvJhT3AiFb_NM73vyQyHjQ

The autopsy revealed it was clearly the vaccine. Obviously,

the CDC disagrees. What was the cause?

The answer is they don't explain any of these things because they can't. And nobody is going to hold them accountable, especially the mainstream media. The mainstream press will not ask any of those questions. It is never going to happen.

"I, James Fuller will say this right here, we are literally at no choice with pharma, elites, oligarchs, and our own governments. Those who can't or won't explain the details to help others do better Ought to lose their positions to someone who can."

One unusual side note is the choice of journal for their paper. In February 2021, the pro-vax editor of that journal, John McConnell wrote an op-ed praising the safety of the COVID vaccines[21]. Exactly one year later, the journal announced on February 25, 2022 that he's dead[22]. Did he die after the booster? What did the autopsy show

---

21.   https://www.google.com/
url?q=https%3A%2F%2Femail.mg1.substack.com%2Fc%2FeJwlkcuSrCAMhp-
m2WEhoOiCxWzOa1hcojKDYEHsPv32g9NVVIBcKsn3O4Ow5fLWZ65IbrPg-
wSd4FUjIEIhV4WyBK9lPwyKq554zRR3ypJQl7UAHCZETc7LxuAMhpzu7EFOYpBk1z
Ctk5p6b6WYhZrdOFsuHOODnadJrOzT1Fw-
QHKg4QnlnROQqHfEsz7E14P_a6f6rUDNV3FQuwLxf-
fy0fwrGLxapD33_KLVrEBNAfo0zoUElZrNhFSRuvwMnvYzbb-
w7Vjpd94TPZzLKUGMFHzAXGg0bQ5sWSu4ts1VSdCccc4Em_pZjkx1vAOwyoKxoh9
WJT3rYolOHTk9JDu2vquXrWjczz0kKfpl1nMXogW3m9aftwFb2n1cKeB7gWRsBK-
xXEDwI8cf2WWDBKXJ5BeDuh_lqMZJqYkz-UF3SzPPcm7ASWvrc6tKumIj-
RNKdfsvzEumoQ&sa=D&sntz=1&usg=AFQjCNGj2PgQEuMX_IvveI5cXllpWglhew

22.   https://www.google.com/
url?q=https%3A%2F%2Femail.mg1.substack.com%2Fc%2FeJwlkLuOxCAMRb9mKCNeC
aGg2GarLbePeDgJmgQicHaUv18yIyFjg-H6Hm8Rllwuc-SK5A4TXgeYBK-
6ASIUclYoUwxGsr5XXDESDFXcK0dineYCsNu4GXKcboveYszp7u7lKHpJVkMH5oXn
TFMqh8Cc8yPlIISW89zPoD-

was his cause of death? We are not allowed to know that. Nobody is talking. You got to love the transparency.

The silence is deafening

The silence on these issues and other evidence that is counter-narrative is deafening.

This just shows you how corrupt the systems are that so few people are speaking out publicly and showing their outrage like Malone did in the video above.

What Malone did is speak up for scientific integrity.

What others did is remain silent.

All of these institutions and individuals are keeping their mouths shut:

- All members of the CDC and FDA outside committees; they don't want to get kicked off the committees.

- Top universities that are supposed to be supporting scientific integrity like Harvard, MIT, Stanford, UCSF, ... There is nothing from the leaders of these institutions or any of the faculty members.

- Top medical societies like the AMA and IDSA and all others

- Top medical journals: not a single one spoke out about this.

i9gwRkgcDf1CunIBsZkU86kN8Pfh3W_iK9zidz3urflf4sa0fW17R4llbwqRWnN06SsteaT
1SEg2nnFNBR6blQFXHOwCnHFgnWD8rGWi3lc2rPaeHpPvCunq69qV_3lKkmJedj1
WIdrncTt-nzezU9v1MEa8JknUbBIPlBIIflG8q0wIJSkMcJouGDW2wYVRq5FR-
bN9YtZa6wSJNNuT2KpmKjcIzlurXf8GviTk&sa=D&sntz=1&usg=AFQjCNEIAhzENZ
Dopvftly9VZGQZlo-DHA

- Top medical thought leaders like Eric Topol, Monica Gandhi, ...

- Congress

- The mainstream news media

- Even so-called "truth seekers" like Debunk The Funk, ZdoggMD, and Your Local Epidemiologist (Dr. Katelyn Jetelina) were silent on the matter. Amazing!

I don't think any of these institutions or individuals are going to object anytime soon due to fear of retribution.

So, the CDC can continue to withhold all unfavorable data, continue to "study" any deaths that were clearly caused by the vaccine, and everyone will continue to ignore all the safety data that is released that is in plain sight (like the Connecticut study) that is contrary to the narrative.

When will this end?

For scientific integrity to be restored, we need to have more than just Robert Malone, Peter McCullough, Paul Alexander, and several others who are outraged and not afraid to speak out.

We need people who have the courage to be on the right side of history.

I don't know when that will happen.

The pace of adoption seems pretty slow. End of Steve Kirsch comments.

I'll tell you how money is the dirty secret when it comes to living and life on new terms and new rules. You do not have the right to question anything the government supports or does with your tax dollars. Oh, you can gripe about a loved one's death due to a vaccine but that will get little attention today because death is now wholesale and affordable. Think about what I just said. When you put a price on life it becomes nothing, but a commodity as goes business and current preferences. Preferences??? Yes, look at the logic structure, look at the narratives in main media. Your being duped and there is nothing you can do about it or is there?

What happens when this gets out of control with the loss of life now affecting the infectors? That Wuhan Lab is a fictitious mode of operation. Put this way the spike protein in this plandemic release does not even exist as stated in the CDC archives, I challenge anyone to present it. They have no sample !! Therefore, it's not supported by true science no matter how many masks the government says you should wear. The point here is a lie is perpetuated for the benefit of strong-arming science itself and therefore no threat from science means pseudo-science or opinion science can become the truth.. !!

I hope those who read this understand the implications. If you do not understand these implications, I will tell you just a few important ones.

First, your right to choose is taken away. This means you have no right to question anything or anyone practicing such deliberation which could cause you harm or death. Second, your ability to eat drink be clothed have a home or any such possessions will be up to those authorities which monitor your compliance to the rules being deliberated. Money will be in the hands of the governments and the corporations which inked the new laws and deals to live by. Understand now?

Without a vaccine passport you cannot participate in food stamps or buy groceries at a local grocer without that vaccine passport! All in the name of EUA and pandemic fears. Do you think this sounds unfair? The time we have left to avert the power pipeline of the rich and get back to moral living is quickly closing mid 2024 at best. When all the legal battles settle for the mandates portion to logical thinking and the judges are bought off you can then believe the authorities will seize that power with open arms. So how will they do it? How will these control freaks subdue an entire population on this planet? Daniel Pinchbeck has this to say about crypto currencies which will pinch the last bastions of freedom from the hands of all who know little about currency and how it works. I'm putting in my book this understanding because I think Daniel has valid questions for such new currency venues and how they will work. Along the way I will inject my understanding as well. Here we go.

"I am currently rethinking my views on crypto currencies and what they portend for the future. Like many people, I missed easy opportunities to invest in crypto currencies years ago — even small amounts put in Bitcoin or Ethereum back then would have yielded fantastic returns by now. But I never pursued financial gain for its own sake. This is due to an innate idealism and communitarianism that, even now, I don't entirely regret." *I share this ideal as well.*

In fact, I worry that crypto accelerates our trajectory toward a world dominated by *economic relations above all other concerns. (This is a huge statement which projects implications of which bias money into the hands or back to the hands of the rich)* Many people I know who start to invest in crypto spend a huge amount of their time and energy focused on it.

This activity is highly abstract and compulsive. It hardly differs from currency speculation or day-trading. Crypto remains hyper-volatile, so making good returns on it requires tremendous vigilance and rapid response. *One thing here, once a crypto is bought with dollars or their currency you wait and watch but mostly, you're waiting because any crypto will go through many ups and downs. The trick is to maintain patience and build the nest of the same next economic cycle.*

I feel the need to reach a meta-level understanding of what's happening with crypto. I intend to develop one in this newsletter, over a series of essays. My current view is that it is a fascinating, ambiguous, hopeful, and dangerous phenomenon. I understand why Hilary Clinton is concerned, as she recently told Bloomberg News, that crypto threatens to undermine the legitimacy of national governments. *This is no fly by night statement about the dangers of the power of crypto when it became the doctrine of the day in economics. Neither buy nor sell save you are adhering to the pressure of one source of economics, bartering is out.* I also keep asking myself where all this new crypto wealth is coming from, since crypto, in itself, does not create anything of tangible value, such as food or energy.

In general, the rich continue to get richer (particularly the super-rich) while the poor keep getting poorer. Crypto doesn't seem to counter but rather it exacerbates this trend, while creating a small number of new wealth holders. In that sense, the crypto revolution seems primarily to be another mechanism through which mostly male, mostly white, mostly engineering-oriented elites increase their economic control over the rest of the world through financial sorcery. *I disagree with Daniel here it's all those people who seek power and control and it's not just the white male it's the Chinese, Japanese, Russians, Swiss, British, Pope, Brazilians etc., we are locked into thinking we need currencies we don't as will be explained later.* I find it interesting that psychedelics, shamanism, and altered states of consciousness are being associated with crypto currencies—it is as if psychedelic mysticism provides a needed ideological support for this magical new form of money that seems to come from nowhere. Similarly, the traditional institutions of banking were generally designed to look like Greek or Roman temples, often with columns and friezes, to make an association between money and the long-term stability of the Classical world.

When someone buys Bitcoin, what are they actually buying? They are buying rights to a particular expression of a mathematical code. But Bitcoin is open-source and anyone with the technical knowledge can "fork" Bitcoin and run a new copy of the original source code (as has happened with Bitcoin Cash and other forks). On some level, then, the only reason that Bitcoin holds its value is due to social trust. A group of people agreed that this particular iteration of the Bitcoin block chain has value. They convinced others to join them in their faith. Doesn't this make Bitcoin inherently a Ponzi Scheme, since, if faith evaporates and new people stop participating, the value dissolves? In other words, it is still possible that Bitcoin might become worthless.

The Bitcoin faithful would answer this by noting that fiat currency is no different: It is also a social agreement, or a social relation, in the end. While this is true, a national currency is "backed" by all of the institutions of that country—its infrastructure, roads, education system, military, as well as its legal code and historical *gravitas*. This seems to hold much greater significance than ascribing value to a block chain whose source code is, in theory, infinitely replicable. However, the financial elite manipulate money creation for their own benefit. This is something you can't do with Bitcoin, which is designed so that only 21 million Bitcoins can ever be mined (I believe they are past 18 million now).

The Gini Coefficient measures the distribution of wealth across populations in different counties. The Gini Coefficient of Bitcoin, along with other major crypto currencies, is far worse than that of highly unequal countries such as North Korea. The "early adaptors" possess far more of the asset than will ever be available later. These early adaptors have a tremendous, vested interest in convincing everyone else that their speculative asset has worth. This is particularly important because Bitcoin is not useful as a medium of exchange currently very few like Starbucks use bit coin. *Of this statement I disagree, bit coin is accepted many places now such as PayPal and restaurants and growing base extremely fast.* You can't do much with it, so it is marketed as "digital gold."

When I wrote How Soon Is Now, I felt hopeful about the potential of block chain technology—particularly Ethereum, the decentralized "world computer" which continues to have scaling issues—to help us build systemic design solutions to our ecological and geopolitical crisis. Different block chains can function simultaneously as currencies (or stores of value) as well as internal governance systems. These governance systems can be defined in any way, according to the contracts programmed into the code.

In theory, you could design a block chain-based currency that accrues value through ecologically and socially regenerative activity, or a currency that cannot be owned by any individual but continually circulates between communities. You could do pretty much anything imaginable with the underlying transparent ledger technology. But even if you built a block chain based, ecological regenerative community currency that might save the world, you would then have to figure out how to get other people to assist to it. *There* is a case *in study a coin called Pi that is social oriented is mined by phone and service minded I have some of these and they are still mineable at https://minepi.com/JFuller.* So far, the only motivation strong enough to get people to adopt a new technology or crypto currency at a large scale is perceived self-interest. That is changing FAST as seen here with Children's Health Defense noting the following.

Vaccine passports and digital identities — full speed ahead

As Off-Guardian's Kit Knightly noted[23] on March 1, "Covid might be dying, but vaccine passports are still very much alive."

23.	https://off-guardian.org/2022/03/01/who-moving-foward-on-global-vaccine-passport-program/

In late February, Knightly also pointed out that the WHO, ominously, is working on an "international treaty[24] [8]on pandemic prevention and preparedness" that would invest the global health organization with the authority to preempt national sovereignty in the management of future pandemics and health challenges.

In a five-part series[25], Corey Lynn of Corey's Digs outlined many disturbing implications of the push for vaccine passports. Falsely marketed as a "convenience," the "passports" eventually will encompass far more[26] than just vaccination records:

"From education to health records, finances, accounts, travel, contact info, and more, will all be linked to your QR code, along with biometrics and fingerprints, then stored on the Blockchain."

The longer-term aim, said Lynn, is to achieve "full power and control," down to the individual level, of spending, taxation, education, transportation, food, communications and healthcare, among other domains.

As writer Cherie Zaslawsky sees it[27], globalists "seek to enslave humanity worldwide in their long-dreamed-of totalitarian utopia. That's utopia for them — as the ruling class that owns the world and everything in it — and dystopia for We the People."

Knightly's March 1 commentary drew readers' attention to SMART Health Cards[28] — "a covert federal vaccine passport" — rolled out [29]in roughly half the country thus far, including in states that previously had paid lip service to banning[30] vaccine passports.

---

24.   https://off-guardian.org/2022/02/26/who-planning-new-pandemic-treaty-for-2024/

25.   https://www.coreysdigs.com/?s=vaccine+id+passports

26.   https://www.coreysdigs.com/technology/the-global-landscape-on-vaccine-id-passports-part-2-how-your-digital-identity-is-moving-to-the-blockchain-for-full-control-over-humans/

27.   https://newswithviews.com/vladimir-putin-russias-trump/

28.   https://smarthealth.cards/en/faq.html#What-are-SMART-Health-Cards

29.   https://childrenshealthdefense.org/defender/us-developing-vaccine-passport-system/

Overseen by the Vaccine Credential Initiative[31] (VCI), SMART Health Cards are intended to "issue, share, and validate vaccination records bound to an individual identity[32]" as well as store "other vital medical data."

A late February article in Forbes boasted that more than 200 million Americans[33] can already "download, print or store their vaccination records as a QR code."

VCI was created by the federally funded MITRE Corporation (an MIT spin-off), which receives an estimated $2 billion a year from U.S. taxpayers to develop advanced surveillance technology[34], among other dubious national security pursuits.

MITRE received a $16.3 million CDC contract "to help construct an efficient game plan for the country during the health crisis," and also spearheaded U.S. Department of Homeland Security efforts to "coordinate" responses among the nation's mayors and governors.

Members of VCI's public-private coalition include Amazon Web Services, Microsoft, Oracle, Salesforce, the Mayo Clinic, and the California and New York state governments, as well as "other health and tech[35] heavyweights." Additional organizations are

30.   https://www.coreysdigs.com/global/the-global-landscape-on-vaccine-id-passports-and-where-its-headed-part-1/

31.   https://vaccinationcredential.org/

32.   https://vci.org/about

33.   https://www.forbes.com/sites/suzannerowankelleher/2022/02/24/national-vaccine-quietly-rolled-out/?sh=267c1eac6be6

34.   https://www.businessinsider.com/mitre-research-lab-facebook-fingerprint-project-2020-7?op=1

35.   https://www.forbes.com/sites/suzannerowankelleher/2022/02/24/national-vaccine-quietly-rolled-out/?sh=267c1eac6be6

contributing[36] to the initiative as "data aggregators" and "health IT vendors."

As an inner-circle member of VCI, New York State has been in the vanguard in building out a digital identity infrastructure[37] intended to be interoperable[38] (able to exchange or assemble data) "throughout the United States and abroad."

New York's "Digital Identity" policy[39], conveniently updated in July 2020, stipulates that citizens, businesses and government employees who conduct online business with the state must go through an "identity vetting" process that could involve authentication via "smart card" or "biometrics."

Refuse totalitarian tyranny

Almost immediately after the COVID shots began being rolled out, Dr. Mike Yeadon[40], at one time a chief scientist and vice president at Pfizer, began protesting the push to inject children.

Yeadon also denounced vaccine passports, describing the apps as a sly vehicle for implementing "illegal, medical apartheid" and totalitarian tyranny.

In a more recent talk[41], Yeadon emphasized that the QR codes' global interoperability will translate into 24/7 tracking of every person "in that moment, in that spot, down to the individual level."

---

36.   https://smarthealth.cards/en/issuers.html

37.   https://www.coreysdigs.com/technology/the-global-landscape-on-vaccine-id-passports-part-4-blockchained/

38.   https://www.governor.ny.gov/news/governor-hochul-announces-excelsior-pass-first-nation-validate-passes-issued-us-states-based

39.   https://its.ny.gov/sites/default/files/documents/nys-p20-001_digital_identity_policy_5.pdf

40.   https://childrenshealthdefense.org/defender/dr-mike-yeadon-rfk-jr-the-defender-podcast-safety-mrna-vaccine-technology/

41.   https://doctors4covidethics.org/video-replays-d4ce-symposium-iii-session-iii/

To impress upon the public the dangers[42] of allowing a vaccine passport system to take hold, Yeadon described what it would mean to become an "out-person:"

"One example: Your VaxPass pings, instructing you to attend for your 3rd or 4th or 5th booster or variant vaccine. If you don't, your VaxPass will expire & you'll become an out-person, unable to access your own life !!"

Fortunately, the globalists' stark vision is becoming increasingly apparent[43] to many members of the public, who are coming to understand, as Ron Paul said[44], that "authoritarian politicians will always lie to the people to protect and increase their own power."

Mainstream media outlets also have begun openly worrying[45] that "parents have a long memory when it comes to how their children have been treated."

And, although it may not seem like it, governmental decisions "ARE affected[46] by what citizens do or don't do," said Rappaport, arguing that it's no time to "let up on pressure."

The bottom line at this critical juncture is simple — rather than be lulled into complacency (or distraction) by the latest propaganda, just say no and don't comply.

Don't wear a mask. Don't get tested. Don't accept toxic jabs. And don't download any QR codes or any other tools (no matter how "convenient") that allow the build-out[47] of digital tyranny.

---

42.   https://off-guardian.org/2021/05/07/halt-vaccine-passports/

43.   https://blog.nomorefakenews.com/2022/02/25/the-crime-engine-of-the-world-is-throwing-rods-and-smoking/

44.   https://www.lewrockwell.com/2022/03/ron-paul/is-putin-the-new-coronavirus/

45.   https://www.npr.org/2022/03/07/1084563882/many-parents-are-angry-over-covid-policies-they-could-be-key-to-gop-2022-gains

46.   https://blog.nomorefakenews.com/2022/03/01/everything-is-yesterdays-news-except-war/

47.   https://home.solari.com/control-is-one-person-at-a-time/

This is all good advice, we as a logical society of thinkers must now advise the future in every control aspect, we can muster in order to stifle the literal killing fields being presented in Nature as well as our survival as a species. Put this way a surgical mask would be good for a 0.05-micron airborne particle. The Virus is 0.026 micron in size. This is like a fly going through a chicken cage, where is the logic?

The integration of currency —value exchange and value storage—and governance in different block chains makes clear what we see in our world in practice: Political and economic systems are never independent from each other. Together they make up one political-economic system. For example, the US government and military are deeply interwoven with the fossil fuel industry, which built the world's largest infrastructure over the last two centuries. The US dollar maintains its status as the global reserve currency because it is used as the "petrol dollar," exchanged for fuel across the Middle East. This is also why we can't stop drilling, fracking, and pipelining.

What is happening in crypto is no different in how control mechanisms are being placed which steer the wealth of the system. Individualist industrialists are a hard bunch to change. I suppose even the destruction of the planet would not stop such destructive behavior until its staring them straight in the face. As I mentioned in the introduction, we now have 2755 billionaires in just over twenty years. This is the real insanity of growth economics. You might ask how do you support nearly 3000 billionaires? The answer is simple, you don't, the time will arrive for payment of such thinking and consumption. It's quite clear we as a civilization are *not* ready for the future. We purport to love and hold to without change to the root mechanism which causes its failure. This illogical stand in our civilization is what will ultimately make it fall into disrepair and collapse completely. In my opinion the marker for change to sustainable systems has passed four years ago 2017. Growth economics and Nobel Prizes for such illogical models is why I say this modeling is encouraged. All the leaders agree growth economics is the way forward. I couldn't disagree more !!

This next case in point shows how brutal the thinking of life as non-important and death is acceptable.

A Freedom of Information Act (FOIA) request made by the Public Health and Medical Professionals for Transparency[48] [9]group has revealed that Pfizer was aware of 1,223 possible vaccine related deaths within the first 90 days of their COVID vaccine rollout. *In 2000 you would have been arrested for malpractice and sent to prison for many years. Today your death means nothing to the industry.*

The Public Health and Medical Professionals for Transparency group is comprised of approximately 30 professors and scientists who are passionate about medical transparency – something we shouldn't have to fight for.

Pfizer's documents[49] accounts 42,086 reactions to their COVID-19 vaccine over a 90-day period from December 1st. 2020 – February 28th, 2021. 25,379 reactions were 'medically confirmed' while 16,707 were 'not-medically confirmed.' Pfizer also redacted the total number of vaccine doses that had been administered up to this point.

According to Pfizer, an event that's not medically confirmed comes as a result of reports that come in without medical records and other necessary documents to make a confirmation. Pfizer also notes that,

"In some reports, clinical information (such as medical history, validation of diagnosis, time from drug use to onset of illness, dose, and use of concomitant drugs) is missing or incomplete, and follow-up information may not be available !!"

---

48.   https://phmpt.org/

49.   https://phmpt.org/wp-content/uploads/2021/11/5.3.6-postmarketing-experience.pdf

"An accumulation of adverse event reports (AERs) does not necessarily indicate that a particular Adverse Event was caused by the drug; rather, the event may be due to an underlying disease or some other factor(s) such as past medical history or concomitant medication."

The documents[50] have also revealed[51] that Pfizer was well aware of tens of thousands of other adverse and serious adverse reactions within months of distribution. This includes 1,403 cases of cardiovascular issues and many cases of vaccine induced heart complications[52].

The FOIA request to obtain these documents was made to the Food and Drug Administration (FDA), who at first did not release them. A court order forced them to comply and begin releasing documents. A joint court order[53] explains that the FDA has proposed to produce 500 pages per month. Based on its calculated number of pages, this would mean that it would complete its production in nearly 55 years, in 2076. This is how long it would take for the public to have clarity on why the FDA approved these vaccines. But now 2022 January the court ruled 7 months.

"Until the entire body of documents provided by Pfizer to the FDA are made available, an appropriate analysis by the independent scientists that are members of Plaintiff is not possible. Would the FDA agree to review and license this product without all the documents? Of course not. These independent, world-renowned scientists should be provided the same forthwith." *The question*

---

50.   https://phmpt.org/wp-content/uploads/2021/11/5.3.6-postmarketing-experience.pdf

51.   https://thepulse.one/2021/11/25/pfizer-was-aware-of-over-50k-serious-covid-vaccine-reactions-with-months-of-distribution/

52.   https://thepulse.one/2021/12/05/researchers-afraid-to-publish-vaccine-heart-inflammation-study-risk-funding-loss-from-big-pharma/

53.   https://fingfx.thomsonreuters.com/gfx/legaldocs/egvbkaeggpq/vaccine%20foia%20status%20report.pdf

becomes whose side is the FDA on? Who does the FDA serve? Who pays the FDA bills? We the people to all those questions. Should the FDA be indicted or at least be examined by a grand jury?

Court Order[54]

This is not a surprise, a systematic review[55] in *PLOS journal* analyzed 28 studies and found that adverse events were less likely to appear in published journal articles than unpublished studies (e.g. industry-held data).

Other concerning data from the Pfizer trials was outlined by whistleblower testimony of Brooke Jackson[56]. An article in the British Medical Journal[57] brought light to her claims that some Pfizer COVID vaccine trial data was falsified. The company she worked for, Ventavia, one of the companies contracted to perform vaccine trials for Pfizer, even tried to discredit her claims as false by saying she did not work on the Pfizer trials. But later released documents showed Ventavia was lying[58].

More documents pertaining to the Pfizer data are set to be released in the coming weeks. Public Health and Medical Professionals for Transparency filed another motion[59] that would

---

54. https://fingfx.thomsonreuters.com/gfx/legaldocs/egvbkaeggpq/vaccine%20foia%20status%20report.pdf

55. https://journals.plos.org/plosmedicine/article/file?id=10.1371/journal.pmed.1002127&type=printable

56. https://maryannedemasi.com/publications/f/fresh-doubts-over-data-integrity-in-pfizer-mrna-trial

57. https://www.bmj.com/content/375/bmj.n2635

58. https://maryannedemasi.com/publications/f/fresh-doubts-over-data-integrity-in-pfizer-mrna-trial

59. https://phmpt.org/wp-content/uploads/2021/11/111521-Second-Joint-Status-Report.pdf

force the FDA to expedite the release of requested documents as well.

Despite the findings outlined in this initial batch of documents, the FDA still thought it was necessary and appropriate to authorize emergency use for these products. In doing so, those who have been, and were injured, by COVID vaccines are not eligible for compensation[60] via the Vaccine Injury Compensation Program.

To date, the program has shielded pharmaceutical companies from liability while payouts to victims come from taxpayer dollars. To date the program has paid more than[61] $4 billion to people and families of vaccine injured people in the United States. Vaccines are liability-free products[62]. *Think about this preceding sentence NO LIABILITY.. !! So death is now just a commodity check !! And this is Planet Earth.*

Since COVID vaccines were deployed, vaccine adverse event reporting systems around the world are recording a record number of injuries. By October 15th, 2021, adverse

reactions to prescription drugs, for example, are extremely underreported, perhaps up to 95 percent.[63]

But what about serious adverse reactions to vaccines? A study[64] published on October 7, 2021 in the Journal Toxicology Reports estimates that underreporting of deaths as a result of the COVID vaccines may have resulted in a number 1000 times less than what the actual number is.

---

60.   https://www.nejm.org/doi/full/10.1056/NEJMp2034438

61.   https://www.nejm.org/doi/full/10.1056/NEJMp2034438

62.   https://thepulse.one/2021/12/03/india-declines-pfizer-moderna-request-for-legal-protection-over-adverse-reactions-to-covid-shots/

63.   https://thepulse.one/2021/10/16/study-up-to-95-percent-of-serious-adverse-reactions-to-drugs-go-under-reported/

64.   https://www.ncbi.nlm.nih.gov/labs/pmc/articles/PMC7581376/#bib0695

They also cite a widely distributed Harvard Pilgrim study[65] published in 2010 which suggested that less than 1 percent of vaccine injuries are reported. This includes serious adverse reactions.

Approximately 50 percent[66] of COVID vaccine injuries reported to the Vaccine Adverse Events Reporting System within the last 30 years are all from COVID shots. As of October 15, 2021, Vaccine Adverse Events Reporting System (VAERS) recorded 122,833 serious adverse events of those 17,128 resulted in death, post administration of COVID vaccines. This includes cases from around the globe, not just the US. You can look up the latest numbers https://medalerts.org

events reported[67] worldwide passed 2,344,240 for COVID vaccines alone in the World Health Organization (WHO) reporting system http://www.vigiaccess.org. Many researchers have also pointed out the potential for underreporting as well. According to multiple studies, it's known that serious adverse reactions to prescription drugs, for example, are extremely underreported, perhaps up to 95 percent.[68]

But what about serious adverse reactions to vaccines? A study[69] published on October 7, 2021 in the Journal Toxicology Reports estimates that underreporting of deaths as a result of the COVID vaccines may have resulted in a number 1000 times less than what the actual number is. They also cite a widely distributed Harvard Pilgrim study[70] published in 2010 which suggested that less than 1

---

65.   https://digital.ahrq.gov/ahrq-funded-projects/electronic-support-public-health-vaccine-adverse-event-reporting-system

66.   https://thepulse.one/2021/11/08/50-of-serious-vaccine-injuries-reported-in-last-30-years-are-from-covid-shots/

67.   http://www.vigiaccess.org/

68.   https://thepulse.one/2021/10/16/study-up-to-95-percent-of-serious-adverse-reactions-to-drugs-go-under-reported/

69.   https://www.ncbi.nlm.nih.gov/labs/pmc/articles/PMC7581376/#bib0695

percent of vaccine injuries are reported. This includes serious adverse reactions. Approximately 50 percent[71] of COVID vaccine injuries reported to the Vaccine Adverse Events Reporting System within the last 30 years are all from COVID shots. As of October 15, 2021, Vaccine Adverse Events Reporting System (VAERS) recorded 122,833 serious adverse events of those 17,128 resulted in death, post administration of COVID vaccines. This includes cases from around the globe, not just the US. You can look up the latest numbers https://www.medalerts.org. [10] We can take this evidence as we see it by choice and proceed to demand reason over money and needless death. I subscribe to the children's health defense and John F Kennedy Jr.'s reasons for lawful venues which set the records straight and stops the coercing of scientist and researchers. You can subscribe to his website here

https://childrenshealthdefense.org/about-us/
membership/?utm_source=salsa&eType=EmailBlastContent&eId=6d
0938-4836-b114-8c0948bc9abf

Now another Super fact which has come forward out of the statistics and evidence pathways in Africa. Africa Has 17.46% of World's Population, Only 3% of World's COVID Deaths. Scientists Want to Know Why!

This last example is quite telling on several levels to life as reality unfolds its truths. First, why do the authorities second guess the

---

70.   https://digital.ahrq.gov/ahrq-funded-projects/electronic-support-public-health-vaccine-adverse-event-reporting-system

71.   https://thepulse.one/2021/11/08/50-of-serious-vaccine-injuries-reported-in-last-30-years-are-from-covid-shots/

evidence-based understandings which show clear pathways exist which avert full blown vaccine need. Why suggest something that does not exist in these regions and say they *will exist*? When you examine the evidence, you see the picture quite clearly. According to data from World Meter[72], the population of Africa makes up 17.46% of the world's population[73]. Yet, AP News reports[74] that the WHO data reveal deaths in Africa are 3% of the global total, while deaths in countries with better health care are much higher, such as 46% in the Americas and 29% in Europe.

Nigeria has the highest population in Africa and the government has recorded just 3,000 deaths in the 200 million people who live in the country. In the U.S., AP News reports [75]there are that many deaths every two or three days.

Some credit early lockdowns with low number of infections Across the world, countries and communities went into lockdown to supposedly help "flatten the curve" and slow the spread of the virus. Lesotho[76], the southernmost landlocked country in the world and surrounded by South Africa, locked down their country and their borders before a single person got sick.

In March 2020, the country declared an emergency, closed the schools and went into a three-week lockdown. In early May, the lockdown was lifted, and the country recorded its first confirmed cases.

72.   https://www.worldometers.info/world-population/africa-population/

73.   https://www.worldometers.info/world-population/

74.   https://apnews.com/article/coronavirus-pandemic-science-health-pandemics-united-nations-fcf28a83c9352a67e50aa2172eb01a2f

75.   https://apnews.com/article/coronavirus-pandemic-science-health-pandemics-united-nations-fcf28a83c9352a67e50aa2172eb01a2f

76.   https://www.bbc.com/news/world-africa-54418613

The <u>BBC reported</u> in October 2020 that in a country of 2 million people they recorded 40 deaths in five months and approximately 1,700 cases.

That number rose to 4,137 [77] cases by January 2021 as citizens from South Africa were crossing the border during the holiday season. In addition to border crossings, the government had been releasing people early from quarantine over cost concerns.

However, experts believe thousands of people had crossed the border illegally because they were unable to afford[78] to pay for their COVID-19 test. In response to the rising number of COVID-19 cases in January, the prime minister imposed a curfew on social venues such as bars and nightclubs at 8 pm.

While the quick action that some countries took to lock down their population may have slowed the spread of the virus in the early days, the SARS-CoV-2 is endemic. This means that it is in the environment and no amount of lockdown will eliminate the virus.

It also means that once lockdowns are opened again, the virus will continue to spread, just like flu and cold viruses. Countries like Lesotho that locked down early have experienced infections after the lockdown was lifted.

The one advantage to flattening the curve and reducing the number of infections early in the pandemic was that doctors could have used the time to improve treatment protocols.

Dr. Vladimir Zelenko[79] and the Front Line COVID-19 Critical Care Alliance[80] are two examples of physicians and physician groups that developed[81] treatment protocols[82] during[83] 2020 that have

---

77.   https://www.theguardian.com/global-development/2021/jan/08/we-cant-cope-lesotho-faces-covid-19-disaster-after-quarantine-failures

78.   https://www.theguardian.com/global-development/2021/jan/08/we-cant-cope-lesotho-faces-covid-19-disaster-after-quarantine-failures

79.   https://vladimirzelenkomd.com/

80.   https://covid19criticalcare.com/

proven to be successful and reduce the number of individuals with severe disease or long-haul symptoms.

Using Google Translate[84], Campbell learned Dr. Haruo Ozaki, chairman of the Tokyo Medical Association, had taken notice of the low number of infections and deaths in Africa where many use ivermectin prophylactically[85] and as the core strategy to treat onchocerciasis[86], a parasitic disease also known as river blindness. More than 99% of people infected live in 31 African countries.

Other medications that are commonly available in Africa have also demonstrated effectiveness against COVID-19. For example, hydroxychloroquine and chloroquine have long been used in the treatment and prevention of malaria[87].

Zelenko has published successful results using hydroxychloroquine against COVID-19.

Malaria is one of the leading causes of death in many developing nations in Africa. The illness is triggered by a parasite carried by an infected female mosquito[88] and characterized by flu-like symptoms.

Delays in treatment increase the severity of the illness and the risk of death. According to the WHO[89], there were 219 million cases of malaria diagnosed in 2017 and 92% of those were in the African region.

Finally, Artemisia annua, also known as sweet wormwood, is an herb used in combination therapies to treat malaria[90]. It was used

---

81.   https://www.ncbi.nlm.nih.gov/labs/pmc/articles/PMC7587171/

82.   https://www.microbiologyresearch.org/content/journal/jmm/10.1099/jmm.0.001250

83.   https://c19ivermectin.com/

84.   https://www.tokyo-np.co.jp/article/123988

85.   https://www.youtube.com/watch?v=E1GF0H9V_1g

86.   https://www.who.int/news-room/fact-sheets/detail/onchocerciasis

87.   https://www.ncbi.nlm.nih.gov/labs/pmc/articles/PMC7476892/

88.   https://www.worldatlas.com/articles/countries-with-the-highest-rates-of-malaria.html

89.   https://www.afro.who.int/health-topics/malaria

in traditional Chinese medicine for more than 2,000 years to treat fever.

Today artemisinin, a metabolite of Artemisia, is the current therapeutic option for malaria. The plant has also been studied since the 2003 SARS outbreak for the treatment[91] of coronaviruses[92], with good results.

As the BBC points out[93], the average age in most African countries is much lower than in the rest of the world. Since many who have died are over the age of 80, and the median age in Africa is 19 years, infections are far less likely to result in death.

Only 3% of the population is over age 65 as compared to 16.9%[94] in North America and 19.2%[95] in Europe.

In addition, residential care facilities for the elderly are rare[96] in most African countries.

Weather may also play a part in who gets COVID: Early in the pandemic, researchers from the University of Maryland discovered there was a correlation[97] between[98] the spread of COVID-19 and

---

90.   https://www.ncbi.nlm.nih.gov/labs/pmc/articles/PMC4323188/

91.   https://www.naturalproductsinsider.com/herbs-botanicals/herb-discovered-have-activity-against-sars-cov-2-virus

92.   https://www.nutraingredients-asia.com/Article/2021/01/19/Artemisinin-for-COVID-19-Indian-and-US-firms-launch-clinically-tested-supplement-after-ayurvedic-regulatory-approval

93.   https://www.bbc.com/news/world-africa-54418613

94.   https://www.statista.com/statistics/457822/share-of-old-age-population-in-the-total-us-population/

95.   https://ec.europa.eu/eurostat/cache/infographs/elderly/index.html

96.   https://www.bbc.com/news/world-africa-54418613

97.   https://pubmed.ncbi.nlm.nih.gov/32525550/

98.   https://today.umd.edu/predicting-covid-19-outbreaks-7dd0062a-d9f3-4a6e-aafa-cb5ab866f46a

temperature, humidity and latitude. They found the virus appears to spread better when humidity and temperatures drop.

In addition, temperate weather and sunny skies such as those you see in Africa increase the likelihood that a population will have optimal levels of vitamin D.

Researchers have discovered that maintaining optimal levels of vitamin D reduces the potential[99] for[100] infection[101] and lowers the risk of severe disease.

How does vaccination in Sub-Saharan Africa make sense?

The low numbers of COVID-19 in the sub-Saharan African population continue despite a less than 6% vaccination rate. Meanwhile, *Western nations' vaccination rates are soaring they struggle with rising infection and death rates.*

This is a KEY factor to visualize and put a stamp on. If you are using the supposed scientific method of *Gene therapy mRNA* as is taking place in so called advanced nations and you're seeing the exact opposite results as supposed third world countries then what's the next logical step? You guessed it, STOP the experiment immediately and garner what works and throw out what does not work. Anything else is a lie and waste of time.

With statistics like this, why are Nigerian officials seemingly overlooking the country's low numbers and health status by announcing they would be *stepping up their immunization* schedule, with hopes to give the shot to half the population before February[102] !! Because these officials are now in the *MONEY* stream and all for the experiments as long as they are not taking the poison themselves.

---

99.   https://www.ncbi.nlm.nih.gov/labs/pmc/articles/PMC7965847/

100.   https://www.ncbi.nlm.nih.gov/labs/pmc/articles/PMC7800698/

101.   https://www.ncbi.nlm.nih.gov/labs/pmc/articles/PMC7973108/

102.   https://apnews.com/article/coronavirus-pandemic-science-health-pandemics-united-nations-fcf28a83c9352a67e50aa2172eb01a2f

Put simply, their target is "herd immunity[103]" in a population that is not suffering. Oyewale Tomori is a virologist from Nigeria who sits on several WHO advisory groups. He believes the vaccination level does not need to be as high as it is in the West.

But Abdool Karim, an epidemiologist in South Africa who has advised the government in the past on COVID-19, disagrees. He is calling for an all-out vaccination program to "prepare for the next wave[104]," There was no first wave how can there be a second wave? Remember the reporting agency is the WHO which cried dead chicken coming before the facts where known and unprofessional conduct of the WHO leadership completely undermines the source. The WHO has funding from Bill Gates who is deliberate about reducing population and eugenics.

Abdool says, "Looking at what's happening in Europe, the likelihood of more cases spilling over here is very high." This is highly suspicious speech of which steers a dangerous trail of death to the unsuspecting.

African countries which have had lower rates of infection and deaths since the start of the pandemic vaccinate more and more of their population, one can't help but wonder: Once vaccination programs are underway, will the death rates climb as they have in other[105] areas[106] of the world[107] where vaccination levels are high? Of

---

103.   https://childrenshealthdefense.org/defender/violating-science-who-changes-meaning-herd-immunity/

104.   https://apnews.com/article/coronavirus-pandemic-science-health-pandemics-united-nations-fcf28a83c9352a67e50aa2172eb01a2f

105.   https://www.planet-today.com/2021/08/fully-vaccinated-gibraltar-sees-2500.html

106.   https://www.beckershospitalreview.com/public-health/nearly-60-of-hospitalized-covid-19-patients-in-israel-fully-vaccinated-study-finds.html

107.   https://www.covid-datascience.com/post/israeli-data-how-can-efficacy-vs-severe-disease-be-strong-when-60-of-hospitalized-are-vaccinated

course, they will, but remember the determination of the ones who get the money that's all that matters.

And if the death rates do climb, how will the "health experts" explain the sudden rising number of infections and deaths on a continent that has thus far avoided pandemic levels of COVID-19? They will pseudo-science the narrative saying the *SECOND WAVE HAS ARRIVED!! Before understanding the extent.*

There is one more thing I wish to convey to the reader that is *DELUSION.* Or the power of Delusion. When people have done what they do for a long time over and over, a few generations such is our case and the advent of Amstel Rothschild the architect of the current usury economics you have little chance of coming out in the right place. Profit before caution. In fact, a miracle would have to be performed to convince people the current system is rigged and unsound. In a later chapter I'll explain the psychology of delusion in the economic sense. What is the way out? And who will provide that way?

Each of the manufactures of the Covid vaccines currently available developed and confirmed their vaccines using fetal cell lines, which originated from aborted fetuses.

(https://lozierinstitute.org/an-ethics-assessment-of-covid-19-vaccine-programs/)

For example, each of the currently available Covid vaccines confirmed their vaccine by protein testing using the abortion-derived cell line HEK-293. (https://lozierinstitute.org/an-ethics-assessment-of-covid-19-vaccine-programs/)

Partaking in a vaccine made from aborted fetuses makes me complicit in an action that offends my religious faith. As such, I cannot, in good conscience and in accord with my religious faith, take any such Covid vaccine at this time.

In addition, any coerced medical treatment goes against my religious faith and the right of conscience to control one's own medical treatment, free of coercion or force. Nuremberg Code.

Equally, compelling any employee to take any current Covid-19 vaccine violates federal and state law, and subjects the employer to substantial liability risk, including liability for any injury the employee may suffer from the vaccine.

Many employers have reconsidered issuing such a mandate after more fruitful review with legal counsel, insurance providers, and public opinion advisors of the desires of employees and the consuming public. Even the Kaiser Foundation warned of the legal risk in this respect. (https://www.kff.org/coronavirus-covid-19/issue-brief/key-questions-about-covid-19-vaccine-mandates/)

Three key concerns: First, informed consent is the guiding light of all medicine, in accord with the Nuremberg Code of 1947.

Second, the Americans with Disabilities Act proscribes, punishes and penalizes employers who invasively inquire into their employees' medical status and then treat those employees differently based on their perceived medical status, as the many AIDS related cases of decades ago fully attest; and

Third, international law, Constitutional law, specific statutes and the common law of torts all forbid conditioning access to employment, education or public accommodations upon coerced, invasive medical examinations and treatment, unless the employer can fully provide objective, scientifically validated evidence of the threat from the employee and how no practicable alternative could possible suffice to mitigate such supposed public health threat and still perform the necessary essentials of employment.

As one federal court just recently held, the availability of reasonable accommodations like accounting for prior infection antibody testing, temperature checks, remote work, other forms of testing, and the like suffice to meet any institution's needs in lieu of masks, public shaming, and forced injections of foreign substances into the body that the FDA admits we do not know the long -term effects of.

For instance, the symptomatic can be self-isolated. Hence, requiring vaccinations only addresses one risk: dangerous or deadly transmission, by the asymptomatic or pre-symptomatic employee, in the employment setting. Yet even government official Mr. Fauci admits, as scientific studies affirm, asymptomatic transmission is exceedingly and "very rare." Indeed, initial data suggests the vaccinated are just as, or even much more, likely to transmit the virus as the asymptomatic or pre-symptomatic.

Hence, the vaccine solves nothing. This evidentiary limitation on any employer's decision making, aside from the legal and insurance risks of forcing vaccinations as a term of employment without any accommodation or even exception for the previously infected (and thus better protected), is the reason most employers wisely refuse to mandate the vaccine. This doesn't even address the arbitrary self-limitation of the pool of talent for the employer: why reduce your own talent pool, when many who refuse invasive inquiries or risky treatment may be amongst your most effective, efficient and profitable employees?

This right to refuse forced injections, such as the Covid-19 vaccine, implements the internationally agreed legal requirement of Informed Consent established in the Nuremberg Code of 1947. (http://www.cirp.org/library/ethics/nuremberg/).

As the Nuremberg Code established, every person must "be able to exercise free power of choice, without the intervention of any element of force, fraud, deceit, duress, overreaching, or other ulterior form of constraint or coercion; and should have sufficient knowledge and comprehension of the elements of the subject matter involved as to enable him to make an understanding and enlightened decision" for any medical experimental drug, as the Covid-19 vaccine currently is.

Second, demanding employees divulge their personal medical information invades their protected right to privacy and discriminates against them based on their perceived medical status, in contravention of the Americans with Disabilities Act. (42 USC §12112(a).) Indeed, the ADA prohibits employers from invasive inquiries about their medical status, and that includes questions about diseases and treatments for those diseases, such as vaccines. As the EEOC makes clear, an employer can only ask medical information if the employer can prove the medical information is both job-related and necessary for the business. (https://www.eeoc.gov/laws/guidance/questions-and-answers-enforcement-guidance-disability-related-inquiries-and-medical). An employer that treats an individual employee differently based on that employer's belief the employee's medical condition impairs the employee is discriminating against that employee based on perceived medical status disability, in contravention of the ADA. The employer must have proof that the employer cannot keep the employee, even with reasonable accommodations, before any adverse action can be taken against the employee. If the employer asserts the employee's medical status (such as being unvaccinated against a particular disease) precludes employment, then the employer must prove that the employee possible a "safety hazard" that cannot be reduced with a reasonable accommodation. The employer must prove, with objective, scientifically validated evidence, that the employee poses a materially enhanced risk of serious harm that no reasonable accommodation could mitigate. This requires the employee's medical status cause a substantial risk of serious harm, a risk that cannot be reduced by any another means. This is a high, and difficult burden, for employers to meet. Just look at the all-prior cases concerning HIV and AIDS, when employers discriminated against employees based on their perceived dangerousness, and ended up paying millions in legal fees, damages and fines.

Third, conditioning continued employment upon participating in a medical experiment and demanding disclosure of private, personal medical information, may also create employer liability under other federal and state laws, including HIPAA, FMLA, and applicable state tort law principles, including torts prohibiting and proscribing invasions of privacy and battery. Indeed, any employer mandating a vaccine is liable to their employee for any adverse event suffered by that employee. The CDC records reports of the adverse events already reported to date concerning the current Covid-19 vaccine.   (https://www.cdc.gov/coronavirus/2019-ncov/vaccines/safety/vaers.html)

Finally, forced vaccines constitute a form of battery, and the Supreme Court long made clear "no right is more sacred than the right of every individual to the control of their own person, free from all restraint or interference of others." (https://www.law.cornell.edu/supremecourt/text/141/250)

Chapter 4

Under the Guise of Compensation

The following information is given directly from children's Health defense and follows the funding venues which pharmacies do not pay a cent. This all is put onto the people as another false hope. Think about this a moment. When you read this information and form your understanding of why they need this, then it all makes perfect sense.

The pharmaceutical companies to grow and become large must be protected from _Liability_. Today we have a most atrocious approach to better health and well-being under the guise of compensation if you die !! Oh well the family could profit off your death.

Now we are seeing droves of people taking the shots for the diseases and virus that might hurt you if not treated with the mRNA gene therapy which is totally new being done on humans. When you think about the whole matter of non-liability, wow okay let's make all the stuff we can which _Might mitigate_ a virus but really who cares if it does or does not, we are not liable for the outcome...off to the races we go to hell with humanity as long as I become rich, money rich.. !

This is exactly what is taking place on this planet today tomorrow and the next until the oops moment arrives and these who make the death are engulfed in it themselves. This has already happened once, and it will surely happen again. Those not in the correct and proper health will perish with the false appetite that money brings. This brings us to the surprising reality of knowing _Motive._ The pace of which this proceeds makes all the difference in the world because most people simply go along with the program that's been etched into the conscious and subconscious minds through media and constant dis information. I myself had the idea of a wonder medicine which could cure the world of all ailments, but this type thinking is not accurate nor are they probable at our time in development as a species. We think we know all the answers, but reality tells a whole different story.

We think we are so intelligent that the universe will succumb to our will and do as we say and act. This is a delusion; we cannot speed the pace of efficacy for the _WANT_ of a whim and call it a cure.. !

In 1986, the U.S. Congress passed the National Childhood Vaccine Injury Act of 1986[1], [11] often simply referred to as the Vaccine Act. Under this act, a no-fault program for administering vaccine claims, known as the National Vaccine Injury Compensation Program[2] (VICP) was established.

Through this program, any individual claiming a vaccine injury (or a parent or guardian of a child) can file a petition with the U.S.Court of Federal Claims. The petition is reviewed by the U.S. Department of Health and Human Services (HHS), which makes a preliminary recommendation.

The U.S. Department of Justice (DOJ) then prepares a legal report, which includes the medical recommendation, and submits it to the court. The court then appoints a special master, who may convene a hearing, and who decides whether the petitioner should be compensated, and if so, what the level of compensation will be.

This compensation is then disbursed to the petitioner through HHS. Petitioners may also appeal a decision that isn't in their favor, and by rejecting the decision of the court, may then file a lawsuit in civil court against the vaccine maker and/or the healthcare provider who administered the vaccine.

VICP, however, does not encompass all vaccines. It covers[3] vaccines that are routinely administered to children and to pregnant

---

1.   https://www.congress.gov/bill/99th-congress/house-bill/5546

2.   https://www.hrsa.gov/vaccine-compensation/index.html

women, and that are subject to the previously-mentioned 75-cent excise tax.

To date, more than 8,400 VICP claims have been settled[4], out of more than 24,000 petitions, with a total of $4.6 billion issued in settlements.

Compensation has also been issued. However, most such settlements were reached following negotiations instead of a hearing, with no admission on the part of HHS that vaccines were ultimately responsible for the injuries in question.

A different category of vaccines, including, at present, the existing COVID-19 vaccines, are covered under what is known as the Countermeasures Injury Compensation Program[5] (CICP).

This program was established under the aegis of the Public Readiness and Emergency Preparedness (PREP) Act[6] of 2005. The PREP act was developed to coordinate the response to a "public health emergency." The law is scheduled to remain in place until 2024.

CICP specifically focuses on countermeasures, that is, "a vaccination, medication, device or other item recommended to diagnose, prevent or treat a declared pandemic, epidemic or security threat."

Under CICP, a different claims process[7] exists as compared to the VICP. The process for claimants is more cumbersome, and individuals have only one year after the administration of the vaccine to file a claim. Injuries whose symptoms materialize later in life, for instance, would presumably not be covered under this process.

---

3.   https://www.hrsa.gov/vaccine-compensation/covered-vaccines/index.html

4.   https://www.hrsa.gov/sites/default/files/hrsa/vaccine-compensation/data/ vicpmonthlyreporttemplate%2011-01-21.pdf

5.   https://www.hrsa.gov/cicp

6.   https://www.phe.gov/Preparedness/legal/prepact/Pages/default.aspx

7.   https://www.hrsa.gov/cicp/filing-benefits

Moreover, the likelihood of success, if past precedent is any indication, is slim. As previously reported by The Defender[8]:

"The program's parsimonious administrators have compensated under 4% of petitioners[9] to date — and not a single COVID vaccine injury — despite the fact that physicians, families and injured vaccine recipients have reported more than 600,000 COVID vaccine injuries[10]."

Notably, vaccines with full FDA approval but which are not placed on a vaccination schedule for children or pregnant women are subject to[11] ordinary product liability laws, while vaccines administered under an Emergency Use Authorization[12] are protected from legal liability.

Furthermore, a 2011 Supreme Court decision[13], Bruesewitz v. Wyeth, held that the Vaccine Act preempts claims made under state-designed defect laws, against vaccines covered by the Act. The decision stated[14] that ""[The Vaccine Act] reflects a sensible choice to leave complex epidemiological judgments about vaccine design to the FDA and the National Vaccine Program rather than juries. "Okay its time to Grand Jury the FDA for violations of efficacy determination. !!"

---

8.   https://childrenshealthdefense.org/defender/mainstream-media-fda-approval-pfizer-vaccine/

9.   https://www.hrsa.gov/cicp/cicp-data

10.   https://childrenshealthdefense.org/defender/vaers-cdc-deaths-adverse-events-covid-vaccines-booster-shots-september/

11.   https://childrenshealthdefense.org/defender/judge-allen-winsor-pfizer-eua-comirnaty-vaccines-interchangeable/

12.   https://www.fda.gov/emergency-preparedness-and-response/mcm-legal-regulatory-and-policy-framework/emergency-use-authorization

13.   http://www.supremecourt.gov/opinions/10pdf/09-152.pdf

14.   https://www.policymed.com/2011/03/supreme-court-rules-in-favor-of-protecting-vaccine-makers-from-state-lawsuits.html

Until the 1980s, a series of successful lawsuits against vaccine makers was seen[15] as resulting in increasing vaccine hesitancy and declining vaccination rates, as indicated in a 1985 National Research Council publication[16], released just one year before the passage of the Vaccine Act. By the way the death rate actually went DOWN those years. Who is lying to who?

Canada:

In recent years, Canada was the only[17] G7 country without a nationwide no-fault vaccine injury compensation program. On a provincial level, Quebec established such a program in 1985, at which time calls[18] for the creation of a national program followed. Attempts[19] were made to develop a national program at this time, which ultimately failed.

As of 2018, Quebec's program had approved[20] a total of 43 claims, paying $5.49 million (CAD) in compensation.

In June 2021, launched[21] a national vaccine injury compensation program, the Vaccine Injury Support Program[22]. The program covers all provinces except Quebec, whose provincial program will continue to operate.

---

15.   https://www.historyofvaccines.org/content/articles/vaccine-injury-compensation-programs

16.   https://www.ncbi.nlm.nih.gov/books/NBK216813/

17. https://cmajnews.com/2021/02/05/vaxinjury-1095922/

18. https://pubmed.ncbi.nlm.nih.gov/3756701/

19. https://www.ncbi.nlm.nih.gov/pmc/articles/PMC2796522/

20.   https://www.canada.ca/en/public-health/services/reports-publications/canada-communicable-disease-report-ccdr/monthly-issue/2020-46/issue-9-september-3-2020/vaccine-injury-compensation-programs-quebec.html

21.   https://www.ctvnews.ca/health/coronavirus/canada-launches-its-first-national-vaccine-injury-compensation-program-1.5451579

22.   https://www.vaccineinjurysupport.ca/en

While this program is funded by Public Health Canada, it is administered by a private company, RCGT Consulting.

The program covers[23] claimants who received a Health Canada-authorized vaccine (on or after Dec. 8, 2020), administered in Canada, with a resulting injury that is serious and permanent or which has resulted in death, and which was reported to the healthcare provider that administered the vaccine.

Though it wasn't until a few months ago that Canada was able to establish a nationwide vaccine compensation program, COVID vaccine manufacturers were already, as of December 2020, indemnified[24] against claims of vaccine injuries.

United Kingdom:

In the UK, the Vaccine Damage Payment Scheme[25] (VDPS) provides compensation totaling £120,000 to anyone who suffers a disability of 60% or more, as a result of their vaccination.

The percentage figure refers to a severe disability[26] resulting in such injuries as the loss of a limb, an amputation, losing 60% or more of normal vision or severe narcolepsy.

Additionally, the 1987 Consumer Protection Act[27] also applies to those who have sustained a vaccine injury, if is found[28] that the product in question did not meet safety standards or was defective.

---

23.   https://www.vaccineinjurysupport.ca/en/faq#eligibility

24.   https://globalnews.ca/news/7521148/coronavirus-vaccine-safety-liability-government-anand-pfizer/

25. https://www.gov.uk/vaccine-damage-payment

26.   https://theconversation.com/uk-citizens-get-less-legal-protection-for-covid-jabs-than-other-vaccines-and-that-could-undermine-confidence-151455

27.   https://www.legislation.gov.uk/ukpga/1987/43/contents

28.   https://fullfact.org/health/unlicensed-vaccine-manufacturers-are-immune-some-not-all-civil-liability/

This is further strengthened by the 2005 General Product Safety Regulations[29].

Consumer protection rights still apply for people injured by the COVID vaccine, as the government wasn't allowed to take those away. But due to the legal definition of defects, and a rule known as the state-of-the-art defense, it is difficult to get compensation when specific problems with the vaccine are not yet known.

COVID vaccines have been added to the VDPS. However, according to the Human Medicines Regulation of 2012[30], protection against civil liability is provided to vaccine manufacturers for unlicensed products issued under a temporary use authorization by the Medicines and Healthcare Products Regulatory Agency.

This regulation was further amended by the Human Medicines (Coronavirus and Influenza) (Amendment) Regulations 2020[31], providing extended immunity from civil liability to vaccine makers and those administering vaccinations. However, the consumer protection laws mentioned above still apply.

Legal indemnity has also been directly provided[32] to vaccine manufacturers in the case of the COVID-19 vaccine.

European Union

The UK laws are based largely on EU legislation, which was codified into British law prior to Brexit.

For instance, the UK Human Medicines Regulations of 2012 and 2020 are largely based on their EU equivalent, EU Directive 2001/83/EC[33] relating to medicinal products for human use. This

---

29.   https://www.legislation.gov.uk/uksi/2005/1803/contents/made

30.   https://www.legislation.gov.uk/uksi/2012/1916/regulation/345

31.   https://www.legislation.gov.uk/uksi/2020/1125/contents/made

32.   https://www.independent.co.uk/news/health/coronavirus-pfizer-vaccine-legal-indemnity-safety-ministers-b1765124.html

33.   https://eur-lex.europa.eu/legal-content/en/TXT/?uri=CELEX%3A32001L0083

includes protections against civil actions for products released under temporary or emergency authorizations.

The 1987 Consumer Protection Act in the UK is, in turn, equivalent to the EU's Directive 85/374/ECC[34] of 1985, on the approximation of the laws, regulations and administrative provisions of the Member States concerning liability for defective products, while the 2005 General Product Safety Regulations were harmonized with EU Directive 2001/95/EC[35] on general product safety.

At the EU level, immunity for vaccine manufacturers was not standard prior to COVID, when legal responsibility tended[36] to lie with the companies.

This, however, is not the case with the COVID vaccines. Under pressure[37] from Vaccines Europe, a trade organization representing vaccine manufacturers in the EU, and under the guide of "ensuring access" to vaccines, exemptions[38] from liability were granted to companies such as AstraZeneca.

Notably, a question[39] posed in August to the European Parliament by one of its elected representatives, Ivan Vilibor Sinčić of Croatia, regarding liability for COVID-19 vaccine side effects, remains unanswered as of this writing.

Within the EU, different member states have enacted their own legislation with regard to vaccine injury compensation claims. These

---

34. https://eur-lex.europa.eu/legal-content/EN/TXT/?uri=celex%3A31985L0374

35. https://eur-lex.europa.eu/legal-content/EN/ALL/?uri=celex%3A32001L0095

36. https://www.france24.com/en/live-news/20210420-covid-vaccine-makers-largely-protected-on-side-effects

37. https://www.documentcloud.org/documents/7046985-European-vaccine-memo.html

38. https://uk.reuters.com/article/us-astrazeneca-results-vaccine-liability/astrazeneca-to-be-exempt-from-coronavirus-vaccine-liability-claims-in-most-countries-idUKKCN24V2EN

39. https://www.europarl.europa.eu/doceo/document/P-9-2021-004007_EN.html

programs were summarized in a 2021 study[40] examining such policies on a global basis. They can be summarized as follows:

Austria: The Vaccine Damage Act is a public-law system for the payment of compensation for vaccine injuries by the state. COVID vaccines are included in this program.

Belgium: No vaccine compensation legislation exists.

France: The existing vaccine injury compensation program provides relief only for injuries related to mandatory vaccinations. Claims for injuries resulting from non-compulsory vaccinations fall under the general principles of French civil law. For COVID vaccines, claims can be lodged with the National Office for Compensation of Medical Accidents, without having to prove a defect with the vaccine or fault on the part of healthcare providers.

Germany: A flat-rate no-fault compensation program exists for vaccines that are mandatory or that are publicly recommended, including COVID vaccines.

Greece: A no-fault program doesn't exist, but a May 2021 high court ruling[41] held that those who sustained vaccine injuries are entitled to state compensation.

Italy: A no-fault program providing state compensation for injuries stemming from required or highly recommended vaccines exists, although it is unclear if this extends to COVID vaccines. Claimants are also free to pursue claims under tort law.

Netherlands, Portugal: There is no specific no-fault scheme, but vaccine injury claims can be filed via provisions of the civil code.

Sweden: An insurance fund, Swedish Pharmaceutical Insurance, handles vaccine injury claims out of court. However, new legislation which took effect Dec. 1 will provide additional state compensation for injuries arising from COVID-19 vaccinations.

---

40.   https://www.mdpi.com/2076-393X/9/10/1116/htm#B22-vaccines-09-01116

41.   https://www.ekathimerini.com/news/1160833/court-says-citizens-can-claim-compensation-for-serious-vaccine-injury/

Israel:

In Israel, the Vaccine Injury Compensation Law[42] was passed in 1989, providing compensation to those injured by vaccines, without having to prove negligence.

Earlier this year, COVID-19 vaccines were included under this law.

New Zealand:

New Zealand maintains[43] a no-fault system for accident compensation, including vaccine injuries, under the aegis of the previously-mentioned Accident Compensation Corporation[44] (ACC).

Although most information on claims appears to be classified, financial compensation totaling $1.6 million (NZD) was provided between 2005 and 2019.

The ACC also handles claims[45] related to COVID-19 vaccination.

China:

China's vaccination program differentiates between mandatory and non-mandatory vaccinations, for the purposes of vaccine injury claims.

The 2019 Law on Vaccine Administration[46] establishes a compensation system for deaths or significant injuries, such as organ

---

42. https://ijhpr.biomedcentral.com/articles/10.1186/s13584-021-00490-w

43. https://www.researchgate.net/publication/

   277541912_Compensation_for_Medical_Injury_in_New_Zealand_Does_No-Fault_Increa

   se_the_Level_of_Claims_Making_and_Reduce_Social_and_Clinical_Selectivity

44. https://www.acc.co.nz/assets/oia-responses/vaccine-treatment-injury-data-GOV-003196-

   response.pdf

45.   https://www.odt.co.nz/news/national/acc-pays-out-almost-130000-covid-claims

or tissue damage, stemming from vaccines. Compensation is paid from the vaccination funds of the country's provincial governments.

Draft legislation in 2020 called for mandatory liability insurance[47] for vaccine manufacturers distributing vaccines in mainland China. However, it is unclear if this legislation was enacted.

Japan:

Until recently, Japan did not have a specific no-fault compensation program for vaccine injuries. But temporary programs[48] where the government would provide compensation to vaccine makers for legal claims they sustained due to vaccine injuries had previously been passed in 2009, for the H1N1 vaccine, and again in 2011 until 2016.

However, a 2020 amendment[49] to Japan's Immunization Act now allows the government to take on the liability risks for COVID-19 vaccines.

India:

India has no specific no-fault legislation[50] under the Drugs and Cosmetic Act for injuries stemming from vaccines that are fully licensed by the country's regulator.

---

46.   https://www.loc.gov/item/global-legal-monitor/2019-08-27/
china-vaccine-law-passed/

47.   https://illumineadvisory.com/trends-%26-insights/f/govt-proposes-to-make-vaccine-
liability-insurance-mandatory

48. https://asia.nikkei.com/Business/Pharmaceuticals/Japan-to-shoulder-COVID-vaccine-
liability-risks

49.   https://pj.jiho.jp/article/242986

50. https://indianexpress.com/article/explained/explained-vaccine-makers-and-indemnity-
7374643/

Claimants are, however, able to file claims in consumer courts or in India's High Court, and the country's drug regulator can also take action against vaccine manufacturers for violations of the law.

Indian law does provide[51] for compensation in the event of injury or death following participation in clinical trials.

Notably, the Indian government's negotiations with Pfizer fell through earlier this year when Indian regulators refused[52] to provide it legal protection via indemnity.

Such protection was not provided to the three COVID-19 vaccines which received an emergency use authorization in India: Covishield, Covaxin and Sputnik V.

Adar Poonawalla, the head of the India-based Serum Institute, the world's largest vaccine manufacturer, had previously called[53] for protection from lawsuits for COVID vaccine injuries.

Malaysia and Singapore:

The country has not developed a no-fault vaccination program, unlike[54] nearby Singapore.

Instead, a variety of legal remedies[55] exist for claimants under civil law, including the Sales of Goods Act of 1957, the Consumer Protection Act of 1999, and the Contracts Act of 1950, and under criminal law, including the Poisons Act of 1952 and the Sale of Drugs Act of 1952.

---

51.   https://indianexpress.com/article/explained/explained-vaccine-makers-and-indemnity-7374643/

52.   https://www.indiatoday.in/coronavirus-outbreak/story/indian-vaccine-makers-also-want-indemnity-like-pfizer-1810212-2021-06-03

53.   https://scroll.in/latest/981786/coronavirus-centre-needs-to-protect-vaccine-makers-against-lawsuits-says-serum-institute-chief

54. https://www.malaysiakini.com/columns/562624

55.   https://www.malaysiakini.com/columns/562624

South Africa:

South Africa is another country that did not develop a no-fault vaccine injury compensation fund until recently, but did so as a result of COVID and, apparently, pressure[56] from vaccine manufacturers.

The fund is meant to provide compensation for "serious adverse responses" which lead to "permanent or significant injury, serious harm to a person's health, other damage or death," assuming these injuries were caused by vaccination.

Philippines:

Similar to South Africa, the Philippines only recently[57] set up a no-fault indemnity program, shielding vaccine manufacturers, as well as public officials, from lawsuits, except in instances of gross negligence or willful misconduct.

This same program will also set up a state fund to provide compensation for vaccine injury claims.

Developing world:

Finally, for 92 low- and middle-income countries, the World Health Organization (WHO), along with a private company, Chubb Limited, has begun to administer[58] a no-fault compensation program.

The countries in question are receiving COVID vaccines via the Gavi Alliance's COVAX Advanced Market Commitment[59] (AMC)

---

56. https://mg.co.za/health/2021-04-21-how-south-africas-covid-vaccine-injury-fund-will-work/

57. https://www.scmp.com/week-asia/politics/article/3122955/coronavirus-vaccine-roll-out-philippines-gets-boost-law-give

58. https://www.who.int/news/item/22-02-2021-no-fault-compensation-programme-for-covid-19-vaccines-is-a-world-first

program, with vaccine injury claims[60] processed through the WHO's new program, which is set to remain in effect until June 30, 2022.

No-fault schemes are increasing, but questions remain

With the recent examples of countries such as Canada and Australia, as well as South Africa and the Philippines, developing their own no-fault vaccine injury compensation funds, as well as their further extension to 92 low- and middle-income countries via the WHO, this type of compensation scheme is clearly the predominant method of dealing with financial claims stemming from vaccine injury claims.

As seen in the case of the U.S., such no-fault programs were developed to address claims of increased vaccine hesitancy, as a result of high-profile lawsuits against vaccine makers, and a decline in vaccine production from hesitant pharmaceutical companies which did not want to shoulder the legal and financial risks involved with releasing a new vaccine to the public.

What, however, goes unaddressed in such claims is the vaccine hesitancy, or outright refusals to get vaccinated, as people question why vaccine makers and, in many cases, everyone involved in distributing and administering vaccines, are shielded from legal action.

Such legal shields cast, for many people at least, a net of doubt, calling into question the safety of such vaccines if their manufacturers, distributors, and public health officials involved in their administration feel the need for legal protections. They may wonder why a product that is said to be safe requires such legal shields.

---

59.   https://www.gavi.org/news/media-room/92-low-middle-income-economies-eligible-access-covid-19-vaccines-gavi-covax-amc

60.   https://childrenshealthdefense.org/defender/fda-licensing-pfizer-comirnaty-covid-vaccine/

Such doubts further increase when governments and their agencies, which are essentially acting as guarantors of these vaccines through various no-fault schemes, redact critical information about these products, including their ingredients, and claims[61] that releasing such documentation will take several decades, as the FDA did recently regarding its documents related to the Pfizer-BioNTech COVID vaccine.

This is despite the fact that in the 2011 Bruesewitz v. Wyeth decision[62], the U.S. Supreme Court gave considerable latitude to the FDA for, essentially, knowing better than judges and juries, or state lawmakers, how to regulate vaccines.

Despite this legal shielding, plenty of coverage of adverse reactions[63], and even deaths[64], following vaccinations is making its way into the media, and to the public consciousness, seemingly negating yet another argument in favor of indemnity.

Furthermore, as many no-fault schemes place the burden on taxpayers and government coffers, these financial costs are ultimately borne by the public.

Arguments that claim shielding vaccine makers from lawsuits also helps to keep the cost of these products down can be called into question on such grounds, especially if the government is the one making deals with vaccine manufacturers and paying for these vaccines.

61.　　https://childrenshealthdefense.org/defender/fda-licensing-pfizer-comirnaty-covid-vaccine/

62.　　http://www.supremecourt.gov/opinions/10pdf/09-152.pdf

63.　　https://childrenshealthdefense.org/defender/vaers-cdc-omicron-vaccine-makers-stock-adverse-events-deaths/

64.　　https://childrenshealthdefense.org/defender/jessica-berg-wilson-dies-covid-vaccine-twitter-censors-obituary/

Costs may be reduced in their purchase price, but the same government and same funds are then used to settle vaccine injury claims.

Such claims from vaccine makers, such as Pfizer for instance, also appear to be disingenuous when considering their high marketing budgets[65], which in the U.S., far exceed their research and innovation expenditures.

Arguments can be made that such funding could be redirected towards legal claims, towards reducing vaccine and drug prices, or both.

So, we see the mode of operation and the motive for such operations. Now I would like to ask the people if they enjoy giving their money to the authorities which claim the pathway to safe wellbeing future is by _Gene Therapy mRNA modality_ which destroy lives and is done by happen chance? It's well documented that the efficacy these authorities claim is half or less than the claimed efficacy. Remember these people are out for a buck at your expense, your fears, your gullibility, your acceptance.

I have several good friends which took what the pharmaceutical industry calls a vaccine. It's not a vaccine its gene therapy. mRNA is the instruction to future DNA cycles.

I'm not a doctor and have no credential for medical creation. But I am a good reader and understand what I read by meticulous finding and cross reference. When you look at the medical world what it's become its own language and such. Keeping to the meaning of that language is what I do in understanding such language.

At what point is the language gone from science based to pseudo-science or in other words opinion based? Today In 2022 we are witnessing the pseudo-science on all levels to promote what is

---

65.　https://childrenshealthdefense.org/defender/pfizer-vaccination-ads-news-sponsorships-research/

called a cure for virus. There is one thing these promoters are missing that is the science...!

Science is the discovery of natural phenomena by artificial means. When searching for cures you find them in due course. To allow that life due course to be breached in my opinion is very very dangerous because you are dealing with pathogens. What could this pathogen lead to? Annihilation on a huge scale billion losing their lives because pseudo-science and money thought more important than humanity itself.

We are witnessing this mode of operations daily now and passports and such will engulf reality of why in the first place.

How do the people behind mRNA modality strike the bargain of death so easily? For the sake of life is how the medical world was established and perpetuated. But now today near the end of 2021 we see Omicron and Delta variant as a fearful thing because the paid media tout lies. The evidence is simply put on a grand scale of millions of years. Our bodies have dealt with millions of VIRUSES and survived. The notion our immune systems cannot handle a simple virus as COVID which it's know today that our immune systems recognize the covid spike. !! ALL the variants also as described in the tests for immune antibody response to covid vaccine is 3-5 days !!

This is immutable proof our immune systems need no help to recognize even the Wuhan Protein spike its all the same family !!

Moderna's top shareholders are Baillie Gifford & Co. ($1.59 billion), Vanguard Group ($1 billion), Blackrock ($999.1 million) and Flagship Pioneering ($653.7 million).

As Nov. 30, 2021, rolled around early news reports on Omicron sent vaccine makers' stocks soaring, after Moderna and Pfizer said they were rushing to develop vaccines for the new variant.

Moderna's stock rose 20% on the Friday following Thanksgiving — a short trading day — while Pfizer and its vaccine partner BioNTech saw respective gains of 6% and 14%.

No evidence we need a vaccine for Omicron, but Pfizer makes the case, anyway

Global Justice Now accused[66] Big Pharma[67] of being responsible for the emergence of Omicron by gobbling up profits selling vaccines to wealthy countries, while refusing to share patents and making sure low-income countries get access to COVID vaccines.

Tim Bierley, the organization's pharma campaigner, said:

"Pharmaceutical companies knew that grotesque levels of vaccine inequality would create prime conditions for new variants to emerge. They let COVID-19 spread unabated in low and middle-income countries. And now the same pharma execs and shareholders are making a killing from a crisis they helped to create. It's utterly obscene."

But not everyone agrees that failure to vaccinate causes new variants to emerge, or that Omicron is dangerous.

Dr. Angelique Coetzee, who is credited with discovering the Omicron variant, said she believes the variant may help lead to herd immunity[68].

Coetzee, who chairs the South African Medical Association and who has been a general practitioner for the last 33 years, said Omicron symptoms so far appear mild.

Coetzee wrote for The Daily Mail[69]:

---

66.   https://www.commondreams.org/news/2021/12/05/utterly-obscene-just-8-pfizer-and-moderna-investors-became-10-billion-richer-after

67.   https://childrenshealthdefense.org/defender_category/big-pharma/

68.   https://childrenshealthdefense.org/defender/covid-outbreak-vaccinated-patient-herd-immunity-theory/

69.   https://www.dailymail.co.uk/debate/article-10256373/Dr-ANGELIQUE-COETZEE-discovered-Omicron-says-reacting-threat.html

"No one here in South Africa is known to have been hospitalized with the *Omicron* variant, nor is anyone here believed to have fallen seriously ill with it ... The simple truth is: We don't know yet anywhere near enough about Omicron to make such judgments or to impose such policies ... If, as some evidence suggests, Omicron turns out to be a fast-spreading virus with mostly mild symptoms for most of the people who catch it, that would be a useful step on the road to herd immunity."

Early data support[70] Coetzee's observation.

According to CNBC[71], the South African Medical Research Council, in a report released Saturday, said most patients admitted to a hospital in Pretoria who had COVID didn't need supplemental oxygen.

The report also noted that many patients were admitted for other medical reasons and were then found to have COVID. Pfizer CEO Bourla responded[72] to that news by telling the Wall Street Journal:

"I don't think it's good news to have something that spreads fast. Spreads fast means it will be in billions of people and another mutation may come. You don't want that."

This kind of speech is boring and too repetitious of intentions to make more and more poison, so the elite are safe in knowing who is on what side of the fight or war as it really is war.

Though it's not clear whether there's a need for a new shot, Pfizer can develop a vaccine that targets omicron by March 2022, Bourla

---

70.   https://www.cnbc.com/2021/12/06/omicron-covid-variants-risk-profile-starts-to-emerge-with-early-data-.html

71.   https://www.cnbc.com/2021/12/07/pfizer-ceo-says-omicron-appears-milder-but-spreads-faster-and-could-lead-to-more-mutations.html

72.   https://www.cnbc.com/2021/12/07/pfizer-ceo-says-omicron-appears-milder-but-spreads-faster-and-could-lead-to-more-mutations.html

said. So, there you have it all the talk is targeting more vaccine or Gene Therapy mRNA pseudo-science opinion!!

So, here we have targeted a region that has fared very well through the illogical advanced nation's philosophy to vaccinate everything in sight. Here we have the fear mongering money lover bent on starting the *next wave* that generates the next wave of deaths that they will say are not tied to Vaccines. If the world leaders wish to push such pathways that's their choice, but why should illogic be forced on populations for the love of money and what appears to be looking like death waves at will?

When the reality of such illogical thinking blooms into full scale warfare on biological terms what will survive? Cockroaches I'm sure will love all the carcasses that will approach many billions. But why should any of us agree to such terms as biological failures? Is this all that humanity can muster for the distant future? Will the pharmaceutical industry implode under the weight of reality of the lies being told? And a warning for those leaders which force their people into less than good conditions of health may face their own deaths because of it.

One must remember there are many more forces than leaders ever dreamed of or considered. Like Brazils leader Bolsonaro will face when his attack on indigenous people comes to an end with him hanging from the very trees he had cut down. Is there enough power left in the people to overcome the tyranny? Yes.

Chapter 5

What will we as world citizens do?

First ask this question. Why should we have to expect our leaders are failing? Is all this control needed? Who benefits from human suffering?

Crowds of hundreds and thousands of people gathered in front of Australian embassies and consulates across Europe, Asia, and North America in response to a worldwide call for help and to increase the visibility of the harsh and increasingly punitive measures which Australians have been subjected to in the wake of the coronavirus crisis.

Protestors in Finland, Britain, Spain, the Netherlands, Portugal, the U.S., Canada, South Africa, Russia, Germany, France, Ireland, Italy, Mexico, Japan, Romania, Israel, and others all joined together to speak with one voice in support of the freedom of Australians to make their own medical decisions without facing penalties, with many noting that Australia is being used as a kind of test ground for a broader implementation of COVID passport systems and quarantine camps. At the Australian consulate in New York City, hundreds of protestors could be seen gathered, waving Australian flags in solidarity with those Down Under and displaying signs with messages which read: "We are with you" and "How long until this comes to America?"

The group can be heard chanting "free Australia!"

Rallying the crowd, a speaker said fighting for Australia means fighting for Americans and others around the world who are subject to encroaching tyranny.

"Here in New York City, we fight ... for liberals, fight for conservatives, fight for religious people, fight for everybody to have the right to say 'No,'" he added, in reference to the choice to take the abortion-tainted[1] COVID jabs. "The right to choose, that is all that that we are looking for. And in Australia right now, that right to choose has gone. "Following a rousing rally outside the consulate, the demonstrators marched to Times Square from the consulate, chanting "save Australia." LET US NOT FORGET.

https://twitter.com/hashtag/
sosfromaustralia?src=hash&ref_src=twsrc%5Etfw

from New York are marching to Times Square from the Australian consulate where they held a rally... pic.twitter.com/QHPvX26bw9[2]— LUKE2FREEDOM (@L2FTV) December 4, 2021[3]

Portuguese man Alfredo Rodrigues shared a video message explaining that more than 100 people gathered before the Australian embassy in Lisbon, the nation's capital, in answer to the SOS.

These kinds of stories are beginning to gather force and backing all around the world. Books like mine are being prepared to show a reality which is not a conjured fiction or pseudo-science opinion. Its people who *Care and believe in a better future world without tyranny.*

In this chapter I will explore some good we as a human race, a species and a steward for good can propose the simple steps that bring whole nations to understanding. What will that understanding be? How will it be shaped? Who will shape it? And most of all will there be true togetherness out of all the divisions which now threaten reality for the not-too-distant future? As you can tell I'm patient about the future we can embrace when we understand the

---

1. https://www.lifesitenews.com/news/which-covid-19-vaccines-are-connected-to-abortion/

2. https://t.co/QHPvX26bw9

3. https://twitter.com/L2FTV/status/1467269871583084544?ref_src=twsrc%5Etfw

stakes of not acting to make the needed changes which keep us on the balance of good for the future. What kind of changes will vary from region to region, but one thing is true, we all need to breath, eat, stay warm or cool, communicate, and dream of a better future without the constant weight of censorship? We are in a battle and that battle has dealt its first casualties, the needless deaths of covid vaccine recipients. When we explored these cause and effect in the earlier chapters what did you realize? When every subject was backed by evidence-based information how did you feel being the experiment recipient with little choice and coercion as a convincer?

Nuremberg Germany, not so many years ago there were men that would make war on the world to adjudicate their vision of the future. What became of that vision is right before your eyes AGAIN TODAY. Our civilization, what has it become? Or What is it becoming? To define these questions, we must look at the motivations which drive a civilization. First, are we CIVILIZED? The answer to this question if defined by psychology will say no!! We are not yet civilized enough to accept the responsibility of taking care of a planet. Not that the planet needs taking care of, it did just fine before our industrial revolution. So, the question becomes, what compels a civilization to destroy itself? We certainly have no footing to stand on with what we have accomplished in the areas of subjection and conquering even destroying nature. You would think that after the most horrendous wars in our history, world war I and World War II, we would have learned proliferation of death is bad for communication and understanding all sides. Yet this is what we are essentially agreeing to when we rush into some new Nano technology on the world scale. Pathogens, have a way of expressing themselves out of the initial master's controls. Was Wuhan China deliberate? I say it was deliberate and the protein we seek to find is still lost somewhere to truth. But these control freaks will unleash a real virus manufactured to remove decenters and populations just as planned. How do I know this? It's documented since 1954 a control pathway was not only born but practiced on all levels of society. Near the end of this book, I will divulge this document a 45-page CIA indoctrination of which was found quite by accident and released in a two-part form. It is now since been combined and one file with the programing codes and flow charts included.

What implications must we as the people demand to stop? Do we the people even have the force enough to stop this tactical war against the population for the purpose of controlling them? I ask this question because what I've seen and am witness to is the complete removal of logical pathway thinking. This is by design. Getting back to the story we started with Australia and the illogical zone of total control of a people.

"You are nice people, and that is probably the reason you are suffering so much," Rodrigues commented. "Because the people that are doing all these wrongs [to] you are cowards ... because they know you don't react violently and that's why it's so difficult for us to see, to watch the news on how you are being treated in this incredible country that you have built with your blood, sweat, and tears. " Continuing, Rodrigues said that the Portuguese stand with Australia since "your fight is our fight." "Count on us, we count on you. Stand your ground, don't give up ... We are going to win this war, this battle, which is the battle of our lives."

A group from Romania[4] protested outside the Australian consulate in Bucharest, holding signs showing that Romanians have responded to Australia's distress call and calling for an end to the lockdowns and medical tyranny plaguing the island nation. Prominent European politicians spoke out publicly, deriding the cruel and totalitarian regime sweeping across the once flourishing democracy as being in "clear violation of human and citizen rights."

---

4.   https://www.reignitedemocracyaustralia.com.au/romania-stands-with-australia/

Earlier in the year, Polish members of parliament from the pro-freedom Confederation Liberty and Independence Party came together in front of Warsaw's Australian embassy, criticizing the Australian government's use of lockdowns as "totalitarianism" and comparing their radical political tactics to those of North Korea.

Confederation Chairman Jakub Kulesza recognized the brutality of the Australian police force, which he said has taken to "oppress, harass, and attack peaceful citizens by depriving them of their fundamental freedoms and liberties."

In all, Kulesza noted that the panorama of measures in place apparently to stop the spread of the virus have been to no avail, since the country has seen "record increases in these infections," all of which demonstrates that "drastic restrictions do not make sense."

"It's hard to call it anything else but madness," he added.

Christine Anderson, a German Member of the European Parliament, spoke out on December 4 in support of Australia's protest, blasting the Australian government's restrictions as not being "about breaking the fourth wave; it is all about breaking people."

"Australia does not need a 'no COVID strategy.' What Australia needs is a 'no oppression strategy,'" Anderson charged. "I will do whatever I can to make it known to the world that your once-free and liberal democracy has been transformed into a totalitarian regime which tramples on human rights, civil liberties, and the rule of law."

Addressing those who might believe that the restrictions are in the best interests of the public, Anderson explained that "at no point in history have the people forcing others into compliance been the good guys."

"The welfare of humanity has always been the alibi of tyrants," she continued. "Do you not realize that this vaccine does not protect you from COVID? It does, however, protect you from governmental oppression ... for now that is, but don't think for even a second that this is not going to change tomorrow." Standing in support with Australia's fight for freedom, Anderson urged listeners to "stop our governments from transforming our free and democratic societies into totalitarian regimes." "We need to do it now, we need to stand up now, we need to fight this now because ... silence and compliance enable tyrants." "Let's all stand united and make it clear to our governments we will not be silent, we will not comply ... We are in this together."

What did we learn from Nazi Germany? What did we learn from wiping out Natural resource areas that never recovered and is not recovering? What I speak of is the future food sources *soil conditions*. We are desert makers and killing our soils!!

Artificial Fertilizers, big industries which consume billions, Natural resources being poisoned by pesticides fungicides etc. We as a people, we as a force must stand against such ideas of Capitalism gone to tyrants, liars, and stealers of ideals for their own use.

Patent system is nothing but a way for the big players to steal by changing the original idea just enough to sound different and therefore be used for their purpose. Where is our legal standing? How do we avert this tyranny with legal authority? This is how and by banding together we are one powerful force to recon with. But what changes for the need? Not a want a need must be defined for the whole to accept and communicate that to our governments our leaders that this pathway can be changed to better support the distant future for our species.

If we are to overcome the Ills and errors and mistakes. What better time to make such a change evident and produce the plan which includes all people? Now is the exact time in political and biblical terms as well. I back this up not just by saying the bible expresses the failure of man to the outcome but also to the chronology of real time as history has chosen this is who we are by our choices our decisions in leadership roles.

My case in point is the TIME mentioned in Daniel 12:7. The document from the CIA in 1954 corroborates the first-Time chronology of Abraham to Jesus Christ as being 1954 years. This is known by the New Testament Matthew 1:17 the generations of which ended with Joseph 14,14,14 each 14 is generations to certain events that caused 3x14=42 generations to a Time to remembrance or Circle of Time which meets the requirements God has placed on his creation Man to repent and get back to his life saving words, but for what you may ask? Look these people, these large corporations in control now HATE Gods creation and the argument is man will fail his duty. What duty will humankind fail? The duty to stewardship what was given us as a purpose back to truth. But not all will fail and that is what these freaks are after the soul of the righteous they seek total dominance because this is what they think is their purpose!! Now, we have a motive and a cause. We are given all the instruction we need with what Jesus delivered as good to others as you would to yourself, but the message gets corrupted. The four horse riders a white horse, a red horse, a black horse, a pale horse which combines the three into one. What color comes out of white, red, black? Pale! We all know we are sinners, but do we all repent and accept our personal truth to seek forgiveness through Jesus Christ? Who is this Jesus Christ? He is the expression of God's love for his creation to get back to him by choice. The spirit falls on the *Just* and the *Unjust.. !* this establishes choice !! Not take away choice like Evil does today.

I will explain this fully in chapter 10.

Is Evil in Control today? Yes, they are given a season. Make your choice wisely because these are ravening wolves seeking to destroy all that profess Jesus Christ as the ultimate authority and giver of life, he is the tree of life witness to the fall of man in that garden, the Word of God.

Now we have one other case I would like to expose as total control of the narrative being presented before the people. The case of Julian Assange for his telling the truth, backing that truth with evidence and presenting it for the people to see. No one can deny his information is real. So now the BEAST wants his Hyde. Who is the beast, the USA is the beast and is referenced in Revelations chapter 13! Thirteen stars thirteen stripes thirteen colonies. Why do we prosecute those who expose unethical and immoral actions of governments and major corporations?

What does this say about humanity our species? How can we trust governments that want to punish people who tell the truth? What moral standards are these leaders delivering? When they want truth killed? In my opinion it's the connecting the dots which define a new world order that really is not new at all. Turns out the same mode of operation that wants Julian Assange in jail are working to remove our choices as well as the populous carries much of the burden of moral cantor. *If the government military complex is upset about truth being revealed why didn't the steps to secure such information be taken? Because they wanted it revealed and want the public involved to bring about more division.* This is an ages old tactic and blame it on someone else. Can we suppose the future of our species is safe when leaders such as we have today moral less contrivers of which will see the end as a delicious victory? When in fact the delusion of that thinking has overcome their ability turn back and reconsider their final fate, they think they do service.

How do we define the truth for our current civilization if we are not allowed to know the truth or tell the truth? This brings huge implications on the outcome of Julian Assange because it defines where the public stands in the eyes of the government. I take a little bit of news and bring the author https://thepulse.one/author/arjunw/

Because he makes some valid points, I'm including his piece as follows. "This is coming from a country that tried to assassinate him[5].

[13] CIA. The US claims Assange's actions put lives in danger. He is facing a 175-year prison sentence. Despite this fact, Assange has already been subjected to extreme torture in prison in the UK, as explained[6] by Karen Kwiatkowski, Ph.D., a retired USAF lieutenant colonel. It's both a scary and unsettling thought with regards to what Assange might experience in the US if he and his team are not successful in appealing the decision, which they are currently in the process of doing. When the decision was made not to extradite Assange, it was a big win for press freedom. But now that things have taken a turn for the worse, it means that the US government will likely be able to obtain any precedent that would criminalize common newsgathering, reporting and publishing practices. The war on information is already quite strong as it is, with COVID being a great example. Scientists, doctors, academics and journalists who present opinion, information and even strong evidence that calls into question and/or opposes government rhetoric are subjected to extreme censorship and ridicule. *But how far will the war on information go?*

Assange was charged under the *Espionage Act* and the Computer Fraud and Abuse Act, largely for actions rightfully recognized as protected news-gathering practices. *He made public previously classified documents exposing various immoral and unethical actions taken by the US government and major corporations,* like war crimes

---

5.   https://thepulse.one/2021/09/27/the-cia-wanted-to-assassinate-julian-assange-a-new-investigation-reveals/

6.   https://www.lewrockwell.com/2019/05/karen-kwiatkowski/pray-and-weep/

for example, among many others. If Julian is extradited, he will be put on trial in Alexandria, *Virginia, where he stands no chance of a fair trial. Think about this, If Julian you or any Journalist is tried for doing the job of journalism, making known truths.. !! Then you are put on trial for truth in a court of Wolves and the wolves are hungry* It is where US intelligence agencies are headquartered. The court complex is 15 miles from CIA headquarters. The state is populated by employees of the very sector whose abuses and crimes Julian exposed. The Espionage Act prevents Julian from arguing why he published what he published, what he exposed, and the fact it didn't result in any physical harm. The Espionage Act was originally intended for use against spies. But it's been used against journalists and whistleblowers in recent decades. These new charges against Assange threaten to criminalize reporting in the United States and around the world. This is classic case of the Devil in the details.

Julian Assange Fiancee @stellamoris1[7]: "Julian has been detained in one form or another for 11 years... every time we have a hearing we hear more about the criminal nature of this case" #HumanRightsDay[8] pic.twitter.com/LA8rZU5TNw[9]

— WikiLeaks (@wikileaks) December 10, 2021[10]

A favorite quote of mine when referring to what's happening with Assange comes from Nils Melzer[11], Human Rights Chair of the Geneva Academy of International Humanitarian Law and has served as UN Reporter on Torture and Other Inhumane or Degrading Treatment or Punishment. How far have we sunk if telling the truth

---

7.   https://twitter.com/StellaMoris1?ref_src=twsrc%5Etfw

8.   https://twitter.com/hashtag/HumanRightsDay?src=hash&ref_src=twsrc%5Etfw

9.   https://t.co/LA8rZU5TNw

10. https://twitter.com/wikileaks/status/1469317128977326093?ref_src=twsrc%5Etfw

11. https://www.instagram.com/p/B_xaDQJHjxt/

becomes a crime? How far have we sunk if we prosecute people that expose war crimes for exposing war crimes? How far have we sunk when we no longer prosecute our own war criminals? Because we identify more with them, than we identify with the people that expose these crimes. What does that say about us and about our governments? In a democracy, the power does not belong to the government, but to the people. But the people must claim it. Secrecy disempowers the people because it prevents them from exercising democratic control, which is precisely why governments want secrecy.

The charges laid against Assange have been met with international condemnation from various civil liberties, human rights, and journalistic communities. More than two dozen organizations that include Human Rights Watch, Amnesty International, the Committee to Protect Journalists and more have repeatedly urged the US Department of Justice to drop its prosecution of Assange. I Agree here 100% because if they proceed this power will be wide open to be nullified by the people.

For the first time in the history of our country, the government has brought criminal charges against a publisher for the publication of truthful information...It establishes a dangerous precedent that can be used to target all news organizations that hold the government accountable by publishing its secrets. And it's equally dangerous for U.S. journalists who uncover the secrets of other nations.

Ben Wizner, Director of the American Civil Liberties Union's Speech, Privacy, and Technology Project, American Civil Liberties Union[12]:

Proponents of Assange's extradition would argue that he threatened national security. I would argue, as would many others, that national security has become an umbrella tool to censor

---

12.   *https://www.aclu.org/press-releases/aclu-comment-julian-assange-indictment*

information that exposes unethical and immoral actions of corporations and governments.

We've seen this more within the past few years. For example, Daniel Hale, a former U.S. intelligence analyst was arrested and sentenced to 45 months in prison for violating the Espionage Act. Hale leaked documents [13] [13] about the secretive U.S. drone program, showing 90% of people killed in Afghanistan were innocent bystanders. What does this say about the government's intentions to silence those who expose truth?

Truth that threatens the power & control governments and corporations continue to grow day by day. The most treacherous acts are always done in the name of the most noble cause. What we see today is the *justification of immoral and unethical actions* that are deemed by government to be justified for the good of the world. This requires massive propaganda campaigns to convince a large portion of the citizenry to obey and stigmatize those who don't. The loss of freedom of speech and censorship continues to grow at an unprecedented level. At this rate, anybody who disobeys government mandates, orders or even questions it may, in the future, be deemed a terrorist and/or a threat. But we can change this. You can help out Julian and his team here[14]. https://www.crowdjustice.com/case/assangeappeal.

---

13.   https://thepulse.one/2021/08/02/whistleblower-jailed-for-showing-90-of-people-killed-by-u-s-drones-are-bystanders/

14.   https://www.crowdjustice.com/case/assangeappeal/

Chapter 6

Planetary Health Stewardship Pathways

To find humanities footing we need to understand the implications of our current decisions being made by those we are supposed to trust. When we look in this area, we find quite upsetting pathways which favor wants over needs. I will attempt to bring the facts front and center as usual my desire is to make the crooked pathway a straight pathway.

Remember when Donald Trump talked about nothing wrong in the environment, ecology and such things as that type of narrative of wants. It's those kinds of people guiding our futures which all they see is money. Money itself is not a bad proposition, It's the *love of money* that creates the implications which destroy logical good pathways and narratives and lives to building better futures without coercion that must be used for these corporations to safeguard their existence. We are on the death bed pathway with white old farts.

I will pick on the current administration as much as the past administrations because Mr. Biden is ten times worse for ecology than the Trump administration. Why does Biden see fit to lie about how he relates to ecology and life when his actions show you different and opposite behavior? President Biden was given (3) letters which detail a new technology which solves for resource consumption and no pollution of any kind in energy technology. So you would think wow this is super nice my administration can begin a real solution based action which solved the huge problem of power without resource consumption and non-polluting at that.. !

But what we have seen in return for such good news? Nothing, Nada, Zilch from the very man proclaiming himself king of environmental courage to fix the wrongs.. !! Now he is a laughingstock of the world, now he's not taken seriously because he's not truthful nor actions based to help ecology or environmental matters which establish *correct pathways* and narratives of support. *There is nothing left for us to do but to create the local sensation of the art and go past the roadblock of politics and self-serving acts which President Biden clearly represents.* The man ought to be ashamed of himself for his actions and his pathway of non-conformity, inequality, and mass freedom stealer by trying to mandate *mRNA Gene Therapy* on the masses. This in my opinion is delusional behavior from a leader of the free world. This in my opinion and I still have one and will express it because we all clearly are in trouble if we follow such nonsense. I personally think President Biden ought to resign for his many failures which do cause harm to the whole country. Removing the people as in *We The People* is a violation of our sacred constitution and Democracy which is we the people to form a more perfect union.. !!

President Biden is a total failure in this respect and before more damage can be done should be removed from his office along with Kamila who backs such nonsense and lies of covid19. I've never in my life had to say such about a president of the free world, but today is special because the violations are real and growing. Our founding fathers said rise up if you see wrong, you have a free speech platform and you are more than the tyrant in government.

Let us examine the Debt of the people of the USA.

How Long Can Treasury Pay Federal Bills?

The October 2021 debt limit increase extended Treasury's capacity to pay federal bills into December 2021 and possibly into January 2022. Nonetheless, unless the debt limit is modified, Treasury's cash balances and borrowing capacity at some point will be exhausted. Secretary Yellen warned of dire consequences if the debt limit were not raised before Treasury's resources were exhausted. *OKAY STOP RIGHT THERE.*

What is the cause of these debts? Simple question, but you will receive an answer that is totally off subject and out of place as we the people fund things. Now in reality it's not we fund things as its us being irresponsible with other people's lives and money namely USA citizens. Then Moody's warned that a federal default would deal a "catastrophic blow to the nascent economic recovery from the COVID-19 pandemic" and that "global financial markets and the economy would be upended ... even if resolved quickly." This is an outright LIE. There were added 660 billionaires to the world that's where the money went.

What is the world's fall back currency? The Petro-dollar or USD. This is important to know because it's like the king of the hill you have thousands of other resource and revenue streams which allow a debt to continue. Just ask Donald Trump how that works, he's now an artificial billionaire like all the rest of the artificial value streams created to cause debt. Those running the Federal government should be fired and replaced immediately before the next round of debt incurs great harm to the people and security of the people are imminent if allowed to continue. The President takes oath of office to protect the USA from ALL, repeat ALL threats foreign and domestic. ALL have failed to do this and have put us in a position of compromise. Now for the rest of the story.

Predicting when Treasury's resources would run out is especially hard in 2021. Economic recovery and growth in government revenues is rapid, although subject to "elevated" levels of uncertainty, according to the Federal Reserve. About $1 trillion in COVID-related budgetary resources remains unspent, which may add uncertainty about the pace of federal outlays. Large deficits also push up debt levels. *AGAIN, this is irresponsible with other peoples money. Are the temporary masters in charge going to pay for their mistakes? Not in your life or our children's or theirs. This is hypocrisy to the highest degree. This is where the public calls in the money and makes straight the stealers and liars and takes back the stolen wealth to be used responsibly again as in the founding. But how do we take back? Buy Gold !! Buy Silver. Stop using and consuming for a month and see what happens or two months and you will get their attention.*

The recent infrastructure act (*This is congress agreeing to fix the road problem*) mandated a transfer of $118 billion into the Highway Trust Fund (HTF), which would be held as special Treasury securities. Secretary Yellen stated those securities would be issued on December 15, 2021. A brief (which congressional clients may request)(*Request?? It should be supplied to congress*) from Wrightson/ICAP— a research unit specializing in Federal Reserve and Treasury operations—notes that tax receipts due on that date would bolster Treasury's cash balances. The Congressional Budget Office estimated that Treasury's resources would be "exhausted soon" if HTF securities were issued in mid-December 2021. (*Here you have the failure to be responsible, it's simply shameful and this is our elected officials remember*)

Although Secretary Yellen warned of "scenarios in which Treasury would be left with insufficient resources to continue to finance the operations of the U.S. government" after December 15, 2021, Wrightson/ICAP projects that Treasury's resources would likely—but not certainly—last into the first week of January 2022. *So, who or what translates such debt as OKAY? It's not okay and must be rectified just like an electrical circuit if it blows a fuse. To continue in fashions and narratives which undermine the reason for treasury, the reason for government is to my understanding the highest slap in the face to its people that can be delivered.. !!*

With all these distractions which are on purpose, or these congress and executive branch people simply don't know what they're doing, are piling up the sale of the USA to satisfy debt to the following orders of those who tote the money that the debt created. Treasuries are going out of STYLE!! Why? The US has for several decades when Richard Nixon took us off the gold standard it opened the flood gates of power and control to the wants of the USA. The Petro dollar reigned king, but now it's changing and nations such as BRICS are de coupling their economies away from the USD because the debt is now unmanaged. I don't blame those countries one bit for deciding to take their responsible roads and backing themselves with Gold again a true source of VALUE which can be monetized. What should an ounce of Gold be? $421,000 no way, yes way. Look at the attempt to monetize bitcoin? $42,000 around there and as high as 69,990.90. What is a bitcoin? Its currency backed by industry nothing more nothing less.

In other words, these leaders for the past 45 years have sold debt that the rich get richer the poor get poorer and the stealer of wealth of which the forefathers warned us is here in full bloom. What does this leave the people to form a more perfect union? It leaves them out of democracy, out of the thought stream, and out of the *protections provided by the constitution*. Your freedom of speech is to say nothing, or we will eliminate you. Give me freedom or give me death rings quite real for a USA citizen today. The shameful acts of our leaders are now reaching the heavens and woe unto those leaders for allowing it to happen.

The Debt Limit in Spring of 2022

Updated March, 2022

Debt limit episodes have been a recurrent federal fiscal feature in the past two decades. Since 2002, the debt limit has been modified 19 times. In August 2019, the Bipartisan Budget Act of 2019 (BBA 2019; P.L. 116-37) suspended the debt limit through July 31, 2021.

The limit was reset at just over $28.4 trillion at the beginning of August 2021 and was raised by $480 billion on October 14, 2021. On December 7, 2021, the House voted to amend a Senate bill that would set up an expedited procedure in the Senate to consider a debt limit increase and would delay some cost-saving measures in Medicare and certain other health programs. (*Idiots are not born they are made*)

The debt limit issue in 2021 has a few unique characteristics. The COVID-19 pandemic remains a source of economic uncertainty, but I believe this will end soon because the pathway declared has served its purpose, Don't mess with authority or else. Fiscal responses spurred by the pandemic accelerated the pace of federal debt accumulation. The U.S. Treasury also sharply increased its cash balances in 2020 to accommodate those fiscal responses.

Since 2013, Congress has suspended the debt limit several times. The Bipartisan Budget Acts of 2015 (BBA 2015; P.L. 114-74), 2018 (BBA 2018; P.L. 115-123), and 2019 (BBA 2019; P.L. 116-37) that adjusted statutory caps on discretionary spending imposed by the Budget Control Act of 2011 (BCA; P.L. 112-25) also suspended the debt limit. When those caps expired at the end of FY2021, the need for legislation to modify them was rendered moot. Thus, the usual legislative vehicle for debt limit modifications over the past decade became unavailable in 2021.

Before 2013, debt limit legislation typically specified a set dollar amount on outstanding debt, either in stand-alone debt limit measures or packaged with other provisions, such as appropriations measures.

Extraordinary Measures in Use Since August 2, 2021

The U.S. Treasury has used "extraordinary measures" to help pay federal obligations since August 2, 2021, when Treasury Secretary Janet Yellen declared a "debt issuance suspension period" (DISP).

A DISP *allows Treasury to suspend investments* in Civil Service and U.S. Postal Service retirement funds. (*Why should these hard working people suffer at the hands of imbeciles which neither pay their mistake or responsible to such past and present mistakes only make more trouble) Congress step up and do your jobs. !! make the law.*

Treasury also draws on certain other, smaller funds, such as the Exchange Stabilization Fund. Federal financial operations continue normally, although debt limit restrictions complicate Treasury's debt and cash management. (*To hold these people responsible is now a complication to them*) Secretary Yellen notified Congress that the DISP would be extended on September 28, 2021, October 18, 2021, and November 16, 2021. Once a debt limit episode ends, Treasury must report on its use of extraordinary measures.

Debt Limit Raised $480 Billion in October 2021

On September 21, 2021, the House passed a continuing resolution (H.R. 5305) to fund federal operations through December 3, 2021. The measure also would have suspended the debt limit through December 16, 2022. On September 27, 2021, the Senate declined to close further debate on the bill. Another continuing resolution, enacted on December 3, 2021, provided funding through February 18, 2022, but contained no debt limit measure. (*Why? Because their tired of fighting the debt fight and just giving in to the whims of the debt holders mentality. This is the danger the forefathers talked about, warned about, and here we are right back to the vomit*)

On September 28, 2021, Secretary Yellen wrote Congress that "Treasury is likely to exhaust its extraordinary measures if Congress has not acted to raise or suspend the debt limit by October 18. At that point, we expect Treasury would be left with very limited resources that would be depleted quickly." On September 29, 2021, the House passed a stand-alone measure to suspend the debt limit. (*Instead of acting responsibly to incur the responsible people which sends a signal we are not going to allow such behavior they go along with the charade)* The Senate passed an amended version on October 7, 2021, calling for a $480 billion increase in the limit. After the House deemed to have accepted the revision on October 12, the President signed it on October 14, 2021. About $300 billion of that increase enabled a _reset of extraordinary measures_. So they can star a new cycle of accepted debt again and again and again. Now that is insanity if one ever saw it.

Treasury Cash Balances

Treasury can pay obligations as long as it retains borrowing capacity, cash balances, and funds available through extraordinary measures. In 2020 and 2021, Treasury's cash balances had been much higher than a decade ago. Before the Lehman Brothers investment bank collapsed in September 2008, Treasury cash balances were kept to minimal levels. Balances then were held mostly below $100 billion. A 2015 Treasury advisory committee recommended increasing cash balances (*wonder who was on that advisory? Where they even American's or state citizens?*) to cover an average week's outlays as a precaution against major financial disruptions.

Cash balances rose sharply after the March 2020 COVID-19 pandemic declaration, as then-Treasury Secretary Steven Mnuchin acted to enable rapid disbursement of CARES Act (P.L. 116-136) payments. After the debt limit suspension lapsed at the end of July 2021, Treasury's cash balances had shrunk to $459 billion. Oh my.

In my opinion again, these acts of giving away others problems creating larger problems is on a trajectory of FAILURE which will snap hard at all economics in the world. Everyone wants to be that multi millionaire or billionaire is now way outside the realm of reality. As I said earlier we consume 1.7 earths and we only have one. !!

This kind of pathway hurts all of us over time. This kind of pathway removes support for the health of the planet its people and ecology in general. We are killing soils microbial base, we are killing the seas, oceans, lakes, rivers and sky's. We have had solution based technology for the past Three years and not a single academic, leader, president, or associate would offer to listen or change because it threatens their control and power.

This book in this section is the indictment of all those who will do harm willingly when they know there is a solution and do not back it for the betterment of the human race as a species. This indictment carries weight in the form of the words presented as facts out of science and scientist who understand and want to change for the better. The politics will end, the destroyer of this planet will be imprisoned and the people, the good people will embrace a truthful path void of evil.

Early Sunset of the Employee Retention Credit

Updated December 8, 2021

The Employee Retention Credit (ERC) was designed to help employers retain employees during the Coronavirus Disease 2019 (COVID-19) public health emergency. The Infrastructure Investment and Jobs Act (IIJA, P.L. 117-58) moved the termination date for the credit forward, to September 30, 2021, from December 31, 2021. This change effectively repeals the ERC for the fourth quarter of 2021 for businesses other than recovery startup businesses. Some employers may have anticipated receiving the ERC for the fourth quarter (the IIJA was signed into law on November 15, 2021), and therefore either underpaid their employment tax liability or received an advance refund from the IRS. IRS guidance provides that taxpayers who received ERC advance payments in the fourth quarter of 2021 must repay those amounts. The IRS also provided relief from late deposit penalties for employers that reduced payroll tax deposits in anticipation of receiving the ERC in the fourth quarter of 2021.

So, what is designed to fail is infusions of money that is really a joke by standards which incite a program will be available for small business or the ability to function. Due to a Plandemic which has been nullified as not emergency based at all with true science research as recorded in earlier chapters of this book. When the government can't afford such added expenditures, they add it anyway to show people hey I'm here and creating something for you. It all looks white, clean and neat on the outside, but when the piper come piping you pay the price of their error. I ask is that fair? If not fair, then should it not cease? The federal government is in real trouble for sure and the piper is headed directly in the path of calling in all his music. What gives the government the right to squander people's money again? I'll tell you, ignorance of the people as designed by the very government you support does not support you anymore. They dig bigger ditches. So how do we get off the runaway train?

The answer is so simple it will make you laugh. Becoming a responsible government. Which finds its path by notifying the public what they are planning. Think of it, if they were responsible to everyday decisions they make or don't make why should they even care or bother? Just show up for election and sound really convincing then back to normal ignoring their responsibility. Today there is no plan, there are a lot of zealots which live to force their ideals on everyone else even by taking choice from the commentary as we have today seen on social networks like Facebook, meta, LinkedIn, YouTube, twitter, etc. And a slew of others that censor your beliefs as not relative or relevant to the social prescription of do as I say. I'm not going to go much or too deeply into that social channel because it's really not worth our time. We know what is good for us.

When values like money becomes more important than family and neighborly confidence then we've missed the target and must re-quiver our bow. When industry is overtaking the small business and being left fewer and fewer resource because we must feed the billionaire then we must reconsider our pathway to reality and equality or qualities which meet ecologic needs. If we kill the ecology, we kill ourselves. Sure, the earth is a big place when you're visualizing your immediate field of view. But when you see the science and calculate the rate of decline of the ecologic environment then health a serious implication comes to light. This light I'm talking about is life as it's supposed to be and once was but is no more. Is changed for the sake of consuming wants over responsible waste management, soil care and health, power generation of which we have that meets future needs well into the future 100+ years, non-polluting non resource consuming after setup low-cost low maintenance. This exists today but not used today in 2022. Why?

When we see this visual out of words, is not sufficient to describe the immensity of the true problem, however we can visualize no worms in a garden because artificial fertilizer caused the salting of the soil to become uninhabitable for the worm, microbe and even the life of the soil compromised. This is happening now at a rate of exponential value 4 times what it was 40 years ago. What does that say for the leaders, the academics which should steer the correct direction in education for learning the mistakes of the past. I can see why Elon Musk wants to be on Mars. What's that say for the very intelligent who offer band aids to planetary health requirements Nobel prize or other prizes that go nowhere. *We've entered into the next step of our learning as a species*. What we do together in the next 10 years must be the revamping of existing Money value structures. For example, Time-Equity of Richard Kiernicki comes to mind for such change venues which include ecology. It's obvious we are spellbound, and delusional thinking growth economics will solve our current no care attitude. There is an attitude of self-worth which is incomplete and dangerous. We must care and allow the changes which precipitate all our life values which steer our decisions. Anything less is a conspiracy.. ! Remember, our civilization is crumbling, but the truth attitude will win the day if recognized for what that truth presents.

The value we need is in our choice to help the planet that born us. Certainly, we can become irresponsible, but what does this current pathway ultimately solve? Like a child on a swing enjoying the gravity and motion so too humanity must see it's mistakes and avert the same mistakes which trap it in the delusion of everything is okay. It's three O'clock somewhere and that time advances ever so imperceptibly for each of us, yet we wake one day and say where did all the time go? How did the earth get so messed up? What should be important to us is not money, but TIME as a value. Richard has developed this thinking and I'm including everyone to jump in and think about a time-based equity system.. ! His email richard@richardmkiernicki.com I encourage anyone to find out what it's all about. A hint about the book in chapter 4 he produces a brilliant understanding of value each of us can understand a purpose and way to equality.

We all want equality, right? Wrong, we must weight our life against that which proposes harm usury unfair practices in business and so on. We all want the assurance our leaders will act responsibly, right? So this is what Richard brings up and fashions the links which hold it all together. Hasn't our thinking and intelligence not solved the real problems yet? One look says no on the tally. This is true of our best scientists our best leaders are clueless if they do not act on evidence which presents the future as wanting !! Some know this I'm not picking on the scientists which do their jobs correctly. We can break away from what holds each of us captive if we learn the right pathway narratives. These narratives are not so easily brought forward because of the negative influence of educational purpose instilled into our learning as *"participate in wants instead of participate in the needs."* This is a profound point because the consumer is taught to all of us that we build a bigger castle a bigger business a more luxury vehicle to drive larger farm, larger city. We are missing our purpose as family and responsible actions to nature and ourselves. As I said earlier the calculation of resource use is now 1.7 earths, we have one earth. ! Is that imaginary, a story whipped up?   https://www.google.com/url?sa=t&rct=j&q=&esrc=s&source=web&cd=&cad=rja&uact=8&rankings%2Fbillionaires-by-country&usg=AOvVaw3hgFgoTJ_MK5MLl2-Ixzti

And this link tells the story quickly.

https://www.google.com/url?sa=t&rct=j&q=&esrc=s&source=web&cd=&ved=2ahUKEwjZulEFvwi7

There is also one other thing I wish to point out. When it takes 3 times the resource base to maintain one billionaire and that billionaire is multi billionaire the number of years left to planet earth is cut to 2027.6 all resources will end all life ends shortly after at the current system use with little to no waste management in the works. Basically, what looks like we are doing well is completely opposite when real world calculation consumption is considered with rate of consumption. This is why I'm writing this book to bring forward a reality which cannot be justified as we progress forward with error ridden changes that cannot meet the need of the future system. As I pointed out we have viable mitigation which solve the problem of energy and soil and now capitalism. Will we garner the will to change properly? We are waiting as we have for the past 40 years.

Today we are set on a course of destruction for a purpose not many can see because of the "today we live tomorrow we die" understanding which cannot meet the future need on the wants.

So how do we arrive at the proper conclusion which satisfies the requirements of the future needs? Academics are the first line of defense against improper pathways which are detrimental to our health and futures. Academics must stand up and prescribe the learning points and ideas which establish the right future narrative. We have all the components we need to succeed and make right the past wrongs. We only need to decide when to begin. Richard Kiernicki has set a date for such an event to take place and set a new standard which uses Time Equity for all things real as true value holders. The total transformation of the current monetary system will change over time, but the important thing is that change will incorporate Equality, or will it?

The proportion of the earth to the proportion of industrialization is staggering. The shear abundance and excavation, mining, building, agriculture, deforestation is on the order of unbelievable when viewed as a time consumption change element. To put in perspective the amount of change happening worldwide is akin to conquering an entire galaxy with billions of stars. The rate of change is a single most important factor to consider, because these rates of change are error ridden not well thought out and consuming resource without much consideration as to waste recovery and correct mitigation of such waste.

This in my opinion is the greatest mistake a civilization can't afford. The progress of mankind must be tempered with the checks and balances which afford sustainable mitigation. As I've said a thousand times the wants on this planet are winning the destruction path. We have two major solutions, three if we take on the capitalists. We have Power Technology 1000 times more advanced than today's feeble solar and wind venues. We have Soil remediation Technology which meets the non-polluting requirements of which Dr. Ali Khalvati has so politely solved in his many years of working out what is needed. If our civilization comes to its senses and promote sustainable change, we may have a chance. Nature is a quick healer and capable of huge remarkable response to balance back to a livable planet healthy vibrant people and oceans which become super abundant again. This is not a maybe time anymore, it's now do or destroy.

Chapter 7

Silent Weapons for Quiet Wars

This document represents the doctrine adopted by the Policy Committee of the Bilderburg Group during its first known meeting in 1954. The following document, dated May 1979, was found on July 7, 1986, in an IBM copier that had been purchased at a surplus sale for spare parts.

Silent Weapons for Quiet Wars

http://www.lawfulpath.com/ref/sw4qw/index.shtml[1] - preface
   http://www.lawfulpath.com/ref/sw4qw/

The following document is taken from two sources. The first, was acquired on a website (of which I can't remember the address) listing as its source the book titled Behold A Pale Horse by William Cooper; Light Technology Publishing, 1991. The second source is a crudely copied booklet, which does not contain a copyright notice, or a publisher's name. With the exception of the Forward, the Preface, the main thing that was missing from the first source was the illustrations. As we began comparing the two, we realized that the illustrations, and the accompanying text (also missing from the first) made up a significant part of the document. This has now been restored by The Lawful Path, and so far, as I know, is the only internet copy available complete with the illustrations.

We have no first-hand knowledge that this document is genuine, however many of the concepts contained herein are certainly reasonable, important, and bear strong

consideration. If anyone has additional knowledge about the source of this document; has better copies of the illustrations than the ones posted here; has any missing pieces to this document or

---

1. http://www.lawfulpath.com/ref/sw4qw/index.shtml

has any comments which can improve upon the quality of this document, we will appreciate your comments. The Lawful Path http://www.lawfulpath.com/

Additional information includes confirmation that this policy was adopted by the International "Elites" at the first Bilderberg Meeting in 1954.

Silent Weapons for Quiet Wars

Forward This manuscript was delivered to our offices by an unknown person. We did not steal the document, nor are we involved with any theft from the United States Government, and we did not get the document by way of any dishonest methods. We feel that we are not endangering the "National Security" by reproducing this document, quite the contrary; it has been authenticated and we feel that we are not only within our rights to publish it, but

morally bound to do so. Regarding the training manual, you may have detected that we had to block out the marginal notes made by the selectee at the C.I.A. Training Center, but I can assure you that the manual is authentic and was printed for the purpose of introducing the selectee to the conspiracy. It has been authenticated by four different technical writers for Military Intelligence, one just recently retired who wants very much to have this manual distributed throughout the world, and one who is still employed as an Electronics Engineer by the Federal Government and has access to the entire series of Training Manuals. One was stationed in Hawaii and held the highest security clearance in the Naval Intelligence, and another who is now teaching at a university, and has been working with the Central Intelligence Agency for several years and wants out before the axe falls on the conspirators. We believed that the entire world should know about this plan, so we distributed internationally one hundred of these manuscripts, to ask individuals at top level positions their opinions. The consensus opinion was to distribute this to as many people as who wanted it, to the end that they would not only understand that "War" had been declared against them but would be able to properly identify the true enemy to Humanity.

Delamer Duverus

Silent Weapons for Quiet Wars

Preface

Conspiracy theories are nothing new to history. Plots to "kill Caesar" and overthrow Rome abounded, for instance. However, it is seldom that concrete clues to such plots come to light and are generally known. Silent Weapons for Quiet Wars, An Introduction Programming Manual was uncovered

quite by accident on July 7, 1986, when an employee of Boeing Aircraft Co. purchased a surplus IBM copier for scrap parts at a sale, and discovered inside details of a plan, hatched in the embryonic days of the "Cold War" which called for control of the masses through manipulation of industry, peoples' pastimes, education and political leanings. It called for a quiet revolution, putting brother against brother, and diverting the public's attention from what is really going on. The document you are about to read is real. It is reprinted in its virgin form, with diagrams.

Table of Contents

- Security
- Historical Introduction
- Political Introduction
- Energy
- Descriptive Introduction of the Silent Weapon
- Theoretical Introduction
- General Energy Concepts
- Mr. Rothschild's Energy Discovery
- Apparent Capital as "Paper" Inductor
- Breakthrough
- Application in Economics
- The Economic Model
- Industrial Diagrams
- Three Industrial Classes
- Aggregation
- The E-model
- Economic Inductance
- Inductive Factors to Consider
- Translation
- Time Flow Relationships and Self-destructive Oscillations
- Industry Equivalent Circuits
- Stages of Schematic Simplification
- Generalization
- Final Bill of Goods
- The Technical Coefficients
- The Types of Admittances
- The Household Industry
- Household Models
- Economic Shock Testing
- Introduction to the Theory of Shock Testing

- Example of Shock Testing
- Introduction to Economic Amplifiers
- Short List of Inputs
- Short List of Outputs
- Table of Strategies
- Diversion, the Primary Strategy
- Diversion Summary
- Consent, the Primary Victory
- Amplification Energy Sources
- Logistics
- The Artificial Womb
- The Political Structure of a Nation - Dependency
- Action/Offense
- Responsibility
- Summary
- System Analysis
- The Draft
- Enforcement

TOP SECRET

Silent weapons for quiet wars
Operations Research Technical Manual
TW-SW7905.1

Welcome Aboard

This publication marks the 25th anniversary of the Third World War, called the "Quiet War", being conducted using subjective biological warfare, fought with

"Silent weapons".

This book contains an introductory description of this war, its strategies, and its weaponry.

May 1979 #74-1120

Security

It is patently impossible to discuss social engineering or the automation of a society, i.e., the engineering of social automation systems (silent weapons) on a national or worldwide scale without implying extensive objectives of social control and destruction of human life, i.e., slavery and genocide.

This manual is an analog declaration of intent. Such a writing must be secured from public scrutiny. Otherwise, it might be recognized as a technically formal declaration of domestic war. Furthermore, whenever any person or group of persons in a position of great power and without full knowledge and consent of the public, uses such knowledge and methodologies for economic conquest - it must be understood that a state of domestic warfare exists between said person or group of persons and the public.

The solution of today's problems requires an approach which is ruthlessly candid, with no agonizing over religious, moral or cultural values. You have qualified for this project because of your ability to look at human society with cold objectivity, and yet analyze and discuss your observations and conclusions with others of similar intellectual capacity without the loss of discretion or humility. Such virtues are exercised in your own best interest. Do not deviate from them.

Historical Introduction

Silent weapon technology has evolved from Operations Research (O.R.), a strategic and tactical methodology developed under the

Military Management in England during World War II. The original purpose of Operations Research was to study the strategic and tactical problems of air and land defense with the objective of effective use of limited military resources against foreign enemies (i.e., logistics). It was soon recognized by those in positions of power that the same methods might be useful for totally controlling a society. But better tools were necessary.

Social engineering (the analysis and automation of a society) requires the correlation of great amounts of constantly changing economic information (data), so high-speed computerized data-processing system was necessary which could race ahead of the society and predict when society would arrive for capitulation. Relay computers were to slow, but the electronic computer, invented in 1946 by J. Presper Eckert and John W. Mauchly, filled the bill. The next breakthrough was the development of the simplex method of linear programming in 1947 by the mathematician George B. Dantzig. Then in 1948, the transistor, invented by J. Bardeen, W.H. Brattain, and W. Shockley, promised great expansion of the computer field by reducing space and power requirements. With these three inventions under their direction, those in positions of power strongly suspected that it was possible for them to control the whole world with the push of a button.

Immediately, the Rockefeller Foundation got in on the ground floor by making a four-year grant to Harvard College, funding the Harvard Economic Research Project for the study of the structure of the American Economy. One year later, in 1949, The United States Air Force joined in.

In 1952 the grant period terminated, and a high-level meeting of the Elite was held to determine the next phase of social operations research. The Harvard project had been very fruitful, as is borne out by the publication of some of its results in 1953 suggesting the feasibility of economic (social) engineering. Engineered in the last half of the decade of the 1940's, the new Quiet War machine stood, so to speak, in sparkling gold-plated hardware on the showroom floor by 1954.

With the creation of the maser in 1954, the promise of unlocking unlimited sources of fusion atomic energy from the heavy hydrogen in sea water and the consequent availability of unlimited social power was a possibility only decades away. The combination was irresistible. Although the silent weapons system was nearly exposed 13 years later, the evolution of the new weapon-system has never suffered any major setbacks. This volume marks the 25th anniversary of the beginning of the Quiet War. Already this domestic war has had many victories on many fronts throughout the world.

Political Introduction

In 1954 it was well recognized by those in positions of authority that it was only a matter of time, only a few decades, before the general public would be able to grasp and upset the cradle of power, for the very elements of the new silent-weapon technology were as accessible for a public utopia as they were for providing a private utopia. The issue of primary concern that of dominance, revolved around the subject of the energy sciences.

Energy

Energy is recognized as the key to all activity on earth. Natural science is the study of the sources and control of natural energy, and social science, theoretically expressed as economics, is the study of the sources and control of social energy. Both are bookkeeping systems: mathematics. Therefore, mathematics is the primary energy science. And the bookkeeper can be king if the public can be kept ignorant of the methodology of the bookkeeping.

All science is merely a means to an end. The means is knowledge. The end is control. Beyond this remains only one issue: Who will be the beneficiary? In 1954 this was the issue of primary concern. Although the so-called "moral issues" were raised, in view of the law of natural selection [1] it was agreed that a nation or world of people who will not use their intelligence are no better than animals who do not have intelligence. Such people are beasts of burden and steaks on the table by choice and consent.

Consequently, in the interest of future world order, peace, and tranquility, it was decided to privately wage a quiet war against the American public with an ultimate objective of permanently shifting the natural and social energy (wealth) of the undisciplined and irresponsible many into the hands of the self-disciplined, responsible, and worthy few. In order to implement this objective, it was necessary to create, secure, and apply new weapons which, as it turned out, were a class of weapons so subtle and sophisticated in their principle of operation and public appearance as to earn for themselves the name "silent weapons". ([1) This concept was never more than a questionable theory of elitist Charles Darwin. In conclusion, the objective of economic research, as conducted by the magnates of capital (banking) and the industries of commodities (goods) and services, is the establishment of an economy which is totally predictable and manipulatable. In order to achieve a totally predictable economy, the low-class elements of society must be brought under total control, i.e., must be housebroken, trained, and assigned a yoke and long-term social duties from a very early age, before they have an opportunity to question the propriety of the matter. In order to achieve such conformity, the lower-class family unit must be disintegrated by a process of increasing preoccupation of the parents and the establishment of government-operated day-care centers for the occupationally orphaned children. The quality of education given to the lower class must be of the poorest sort, so that the moat of ignorance isolating the inferior class from the superior class is and remains incomprehensible to the inferior class. With such an initial handicap, even bright lower-class individuals have little if any hope of extricating themselves from their assigned lot in life. This form of slavery is essential to maintain some measure of social order, peace, and tranquility for the ruling upper class.

Descriptive Introduction of the Silent Weapon

What is expected from an ordinary weapon is expected from a silent weapon by its creators, but only in its own manner of functioning. It shoots situations, instead of bullets; propelled by data processing, instead of chemical reaction (explosion); originating from bits of data, instead of grains of gunpowder; from a computer, instead of a gun; operated by a computer programmer, instead of a marksman; under the orders of a banking magnate, instead of a military general. It makes no obvious explosive noises, causes no obvious physical or mental injuries, and does not obviously interfere with anyone's daily social life. Yet it makes an unmistakable "noise," causes unmistakable physical and mental damage, and unmistakably interferes with the daily social life, i.e., unmistakable to a trained

observer, one who knows what to look for.

The public cannot comprehend this weapon, and therefore cannot believe that they are being attacked and subdued by a weapon. The public might instinctively feel that something is wrong, but that is because of the technical nature of the silent weapon, they cannot express their feeling in a rational way, or handle the problem with intelligence. Therefore, they do not know how to cry for help, and do not know how to associate with others to defend themselves against it.

When a silent weapon is applied gradually, the public adjusts/ adapts to its presence and learns to tolerate its encroachment on their lives until the pressure (psychological via economic) becomes too great and they crack up.

Therefore, the silent weapon is a type of biological warfare. It attacks the vitality, options, and mobility of the individuals of a society by knowing, understanding, manipulating, and attacking their sources of natural and social energy, and their physical, mental, and emotional strengths and weaknesses.

Theoretical Introduction

Give me control over a nation's currency, and I care not who makes its laws, Mayer Amshel Rothschild (1743-1812)

Today's silent weapons technology is an outgrowth of a simple idea discovered, succinctly expressed, and effectively applied by the quoted Mr. Mayer Amshel Rothschild. Mr. Rothschild discovered the missing passive component of economic theory known as economic inductance. He, of course, did not think of his discovery in these 20th-century terms, and to be sure, mathematical analysis had to wait for the Second Industrial Revolution, the rise of the theory of mechanics and electronics, and finally, the invention of the electronic computer before it could be effectively applied in the control of the world economy.

General Energy Concepts

In the study of energy systems, there always appears three elementary concepts. These are potential energy, kinetic energy, and energy dissipation. And corresponding to these concepts, there are three idealized, essentially pure physical counterparts called passive components. (1) In the science of physical mechanics, the phenomenon of potential energy is associated with a physical property called elasticity or stiffness and can be represented by a stretched spring. In electronic science, potential energy is stored in a capacitor instead of a spring. This property is called capacitance instead of elasticity or stiffness. (2) In the science of physical mechanics, the phenomenon of kinetic energy is associated with a physical property called inertia or mass and can be represented by a mass or a flywheel in motion. In electronic science, kinetic energy is stored in an inductor (in a magnetic field) instead of a mass. This property is called inductance instead of inertia.

(3) In the science of physical mechanics, the phenomenon of energy dissipation is associated with a physical property called friction or resistance and can be represented by a dashpot or other device which converts energy into heat. In electronic science, dissipation of energy is performed by an element called either a resistor or a conductor, the term "resistor" being the one generally used to describe a more ideal device (e.g., wire) employed to convey electronic energy efficiently from one location to another. The property of a resistance or conductor is measured as either resistance or conductance reciprocals.

In economics these three energy concepts are associated with:

1. Economic Capacitance - Capital (money, stock/inventory, investments in buildings and durables, etc.)

2. Economic Conductance - Goods (production flow coefficients)

3. Economic Inductance - Services (the influence of the population of industry on output)

All the mathematical theory developed in the study of one energy system (e.g., mechanics, electronics, etc.) can be immediately applied in the study of any other energy system (e.g., economics).

Mr. Rothschild's Energy Discovery

What Mr. Rothschild [2] had discovered was the basic principle of power, influence, and control over people as applied to economics. That principle is "when you assume the appearance of power, people soon give it to you."

Mr. Rothschild had discovered that currency or deposit loan accounts had the required appearance of power that could be used to induce people (inductance, with people corresponding to a magnetic field) into surrendering their real wealth in exchange for a promise of greater wealth (instead of real compensation). They would put up real collateral in exchange for a loan of promissory notes. Mr. Rothschild found that he could issue more notes than he had backing for, so long as he had someone's stock of gold as a persuader to show his customers.

Mr. Rothschild loaned his promissory notes to individuals and to governments. These would create overconfidence. Then he would make money scarce, tighten control of the system, and collect the collateral through the obligation of contracts. The cycle was then repeated. These pressures could be used to ignite a war. Then he would control the availability of currency to determine who would win the war. That government which agreed to give him control of its economic system got his support. Collection of debts was guaranteed by economic aid to the enemy of the debtor. The profit derived from this economic methodology made Mr. Rothschild all the more able to expand his wealth. He found that the public greed would allow currency to be printed by government order beyond the limits (inflation) of backing in precious metal or the production of goods and services.

Apparent Capital as "Paper" Inductor

In this structure, credit, presented as a pure element called "currency," has the appearance of capital, but is in effect negative capital. Hence, it has the appearance of service, but is in fact, indebtedness or debt. It is therefore an economic inductance instead of an economic capacitance, and if balanced in no other way, will be balanced by the negation of population (war, genocide). The total goods and services represent real capital called the gross national product, and currency may be printed up to this level and still represent economic capacitance; but currency printed beyond this level is subtractive, represents the introduction of economic inductance, and constitutes notes of indebtedness. War is therefore the balancing of the system by killing the true creditors (the public which we have taught to exchange true value for inflated currency) and falling back on Whatever is left of the resources of nature and regeneration of those resources. Mr. Rothschild had discovered that currency gave him the power to rearrange the economic structure to his own advantage, to shift economic inductance to those economic positions, which would encourage the greatest economic instability and oscillation.

The final key to economic control had to wait until there was

sufficient data and high-Speed computing equipment to keep close watch on the economic oscillations created by price shocking and excess paper energy credits - paper inductance/inflation. Breakthrough The aviation field provided the greatest evolution in economic engineering by way of the mathematical theory of shock testing. In this process, a projectile is fired from an airframe on the ground and the impulse of the recoil is monitored by vibration transducers connected to the airframe and wired to chart recorders.

By studying the echoes or reflections of the recoil impulse in the airframe, it is possible to discover critical vibrations in the structure of the airframe which either vibrations of the engine or Aeolian vibrations of the wings, or a combination of the two, might reinforce resulting in a resonant self-destruction of the airframe in flight as an aircraft.

From the standpoint of engineering, this means that the strengths and weaknesses of the structure of the airframe in terms of vibrational energy can be discovered and manipulated.

Application in Economics

To use this method of airframe shock testing in economic engineering, the prices of commodities are shocked, and the public consumer reaction is monitored. The resulting echoes of the economic shock are interpreted theoretically by computers and the psycho economic structure of the economy is thus discovered. It is by this process that partial differential and difference matrices are discovered that define the family household and make possible its evaluation as an economic industry (dissipative consumer structure). Then the response of the household to future shocks can be predicted and manipulated, and society becomes a well-regulated animal with its reins under the control of a sophisticated computer-regulated social energy bookkeeping system.

Eventually every individual element of the structure comes under computer control through a knowledge of personal preferences, such knowledge guaranteed by computer association of consumer preferences (universal product code, UPC; zebra-striped pricing codes on packages) with identified consumers (identified via association with the use of a credit card and later a permanent "tattooed" body number invisible under normal ambient illumination).

Summary

Economics is only a social extension of a natural energy system. It, also, has its three passive components. Because of the distribution of wealth and the lack of communication and lack of data, this field has been the last energy field for which a knowledge of these three passive components has been developed. Since energy is the key to all activity on the face of the earth, it follows that in order to attain a monopoly of energy, raw materials, goods, and services and to establish a world system of slave labor, it is necessary to have a first strike capability in the field of economics. In order to maintain our position, it is necessary that we have absolute first knowledge of the science of control over all economic factors and the first experience at engineering the world economy.

In order to achieve such sovereignty, we must at least achieve this one end: that the public will not make either the logical or mathematical connection between economics and the other energy sciences or learn to apply such knowledge. This is becoming increasingly difficult to control because more and more businesses are making demands upon their computer programmers to create and apply mathematical models for the management of those businesses. It is only a matter of time before the new breed of private programmer/economists will catch on to the far-reaching implications of the work begun at Harvard in 1948. The speed with which they can communicate their warning to the public will largely depend upon how effective we have been at controlling the media, subverting education, and keeping the public distracted with matters of no real importance.

The Economic Model

Economics, as a social energy science has as a first objective the description of the complex way in which any given unit of resources is used to satisfy some economic want. (Leontief Matrix). This first objective, when it is extended to get the most product from the least or limited resources, comprises that objective of general military and industrial logistics known as Operations Research. (See simplex method of linear programming.) The Harvard Economic Research Project (1948-) was an extension of World War II Operations

Research. Its purpose was to discover the science of controlling an economy: at first the American economy, and then the world economy. It was felt that with sufficient mathematical foundation and data, it would be nearly as easy to predict and control the trend of an economy as to predict and control the trajectory of a projectile. Such has proven to be the case. Moreover, the economy has been transformed into a guided missile on target. The immediate aim of the Harvard project was to discover the economic structure, what forces change that structure, how the behavior of the structure can be predicted, and how it can be manipulated. What was needed was a well-organized knowledge of the mathematical structures and interrelationships of investment, production, distribution, and consumption.

To make a short story of it all, it was discovered that an economy obeyed the same laws all the mathematical theory and practical and computer knowhow developed for the electronic field could be directly applied in the study of economics. This discovery was not openly declared, and its more subtle implications were and are kept a closely guarded secret, for example that in an economic model,

human life is measured in dollars, and that the electric spark generated when opening a switch connected to an active inductor is mathematically analogous to the initiation of war.

The greatest hurdle which theoretical economists faced was the accurate description of the household as an industry. This is a challenge because consumer purchases are a matter of choice which in turn is influenced by income, price, and other economic factors. This hurdle was cleared in an indirect and statistically approximate way by an application of shock testing to determine the current characteristics, called current technical coefficients, of a household industry Finally, because problems in theoretical electronics can be translated very easily into problems of theoretical electronics, and the solution translated back again, it follows that only a book of language translation and concept definition needed to be written for economics. The remainder could be gotten from standard works on mathematics and electronics. This makes the publication of books on advanced economics unnecessary, and greatly simplifies project security.

Industrial Diagrams

The ideal industry is defined as a device which receives value from other industries in several forms and converts them into one specific product for sales and distribution to other industries. It has several inputs and one output. What the public normally thinks of as one industry is really an industrial complex, where several industries under one roof produce one or more products. A pure (single output) industry can be represented over simply by a circuit block as follows:

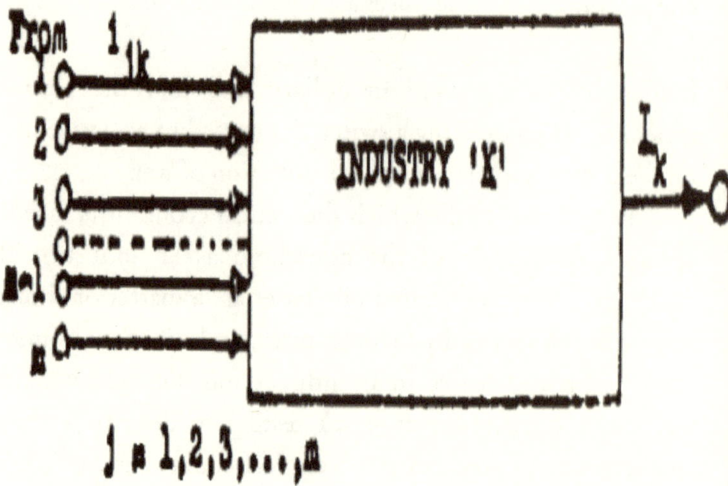

The flow of product from industry #1 (supply) to industry #2 (demand) is denoted by

112. The total flow out of industry "K" is denoted by Ik (sales, etc.).

A three industry network can be diagrammed as follows:

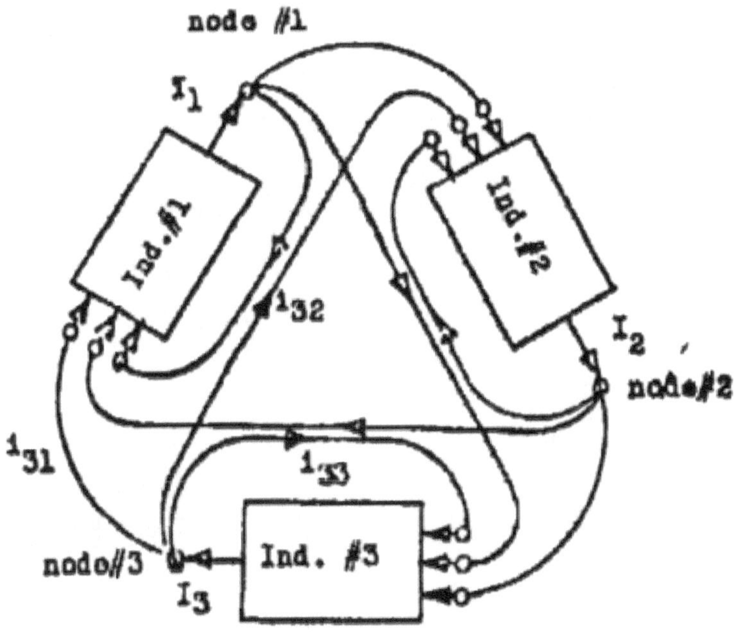

A node is a symbol of collection and distribution of flow. Node #3 receives from industry #3 and distributes to industries #1 and #3. If industry #3 manufactures chairs, then a flow from industry #3 back to industry #3 simply indicates that industry #3 is using part of its own output product, for example, as office furniture. Therefore, the flow may be summarized by the equations:

$$\text{Node \#1}: \quad I_1 = i_{11} + i_{12} + i_{13} = \sum i_{1k}$$

$$\text{Node \#2}: \quad I_2 = i_{21} + i_{22} + i_{23} = \sum i_{2k}$$

$$\text{Node \#3}: \quad I_3 = i_{31} + i_{32} + i_{33} = \sum i_{3k}$$

$$\text{where } \sum \text{ denotes } \sum_{k=1}^{k=3}$$

Three Industrial Classes

Industries fall into three categories or classes by type of output:

    Class #1 - Capital (resources)
    Class #2 - Goods (commodities or use - dissipative)
    Class #3 - Services (action of population

Class #1 industries exist at three levels:

(1) Nature - sources of energy and raw materials.

(2) Government - printing of currency equal to the gross national product (GNP), and extension of currency in excess of GNP.

(3) Banking - loaning of money for interest, and extension (inflation/counterfeiting) of economic value through the deposit loan accounts.

Class #2 industries exist as producers of tangible or consumer (dissipated) products. This sort of activity is usually recognized and labeled by the public as "industry."

Class #3 industries are those which have service rather than a tangible product as their output. These industries are called (1) households, and (2) governments. Their output is human activity of a mechanical sort, and their basis is population.

Aggregation

The whole economic system can be represented by a three-industry model if one allows the names of the outputs to be (1) capital, (2) goods, and (3) services. The problem with this representation is that it would not show the influence, say, the textile industry on the ferrous metal industry. This is because both the textile industry and the ferrous metal industry would be contained within a single classification called the "goods industry" and by this process of combining or aggregating these two industries under one system block they would lose their economic individuality.

The E-Model-A national economy consists of simultaneous flows of production, distribution, consumption, and investment. If all of these elements including labor and human functions are assigned a numerical value in like units of measure, say, 1939 dollars, then to his flow can be further represented by a current flow in an electronic circuit, and its behavior can be predicted and manipulated with useful precision. The three ideal passive energy components of electronics, the capacitor, the resistor, and the inductor correspond to the three ideal passive energy components of economics called the pure industries of capital, goods, and services, respectively:

* Economic capacitance represents the storage of capital in one form or another.

* Economic conductance represents the level of conductance of materials for the

production of goods.

* Economic inductance represents the inertia of economic value in motion. This is a

population phenomenon known as services.

Economic Inductance

An electrical inductor (e.g., a coil or wire) has an electric current as its primary phenomenon and a magnetic field as its secondary phenomenon (inertia). Corresponding to this, an economic inductor has a flow of economic value as its primary phenome non and a population field as its secondary field phenomenon of inertia. When the flow of economic value (e.g., money) diminishes, the human population field collapses in order keep the economic value (money) flowing (extreme case - war). This public inertia is a result of consumer buying habits, expected standard of living, etc., and is generally a phenomenon of self-preservation.

Inductive Factors to Consider

(1) Population

(2) **Magnitude of the economic activities of the government**

(3) The method of financing these government activities (See Peter-Paul Principle - inflation of the currency.)

Translation

(a few examples will be given)

- Charge: Coulombs Dollars (1939)
- Flow/Current: Amperes (coulombs/ second) Dollars of flow per year
- Motivating Force: Volts; Dollars (output) demand
- Conductance: Amperes per volt; Dollars of flow per year per dollar demand

- Capacitance: Coulombs per volt: Dollars of production inventory/ stocks per dollar demand

Time Flow Relationships and Self-Destructive Oscillations

An ideal industry may be symbolized electronically in various ways. The simplest way is to represent a demand by a voltage and a supply by a current. When this is done, the relationship between the two becomes what is called an admittance, which can result from three economic factors: (1) foresight flow, (2) present flow, and (3) hindsight flow.

1. Foresight flow is the result of that property of living entities to cause energy (food) to be stored for a period of low energy (e.g., a winter season). It consists of demands made upon an economic system for that period of low energy (winter season). In a production industry it takes several forms, one of which is known as production

stock or inventory. In electronic symbology this specific industry demand (a pure capital industry) is represented by capacitance and the stock or resource is represented by a stored charge. Satisfaction of an industry demand suffers a lag because of the loading effect of inventory priorities.

2. Present flow ideally involves no delays. It is, so to speak, input today for output today, a "hand to mouth" flow. In electronic symbology, this specific industry demand (a pure us industry) is represented by a conductance which is then a simple economic valve (a dissipative element).

3. Hindsight flow is known as habit or inertia. In electronics this phenomenon is the characteristic of an inductor (economic analog = a pure service industry) in which a current flow (economic analog = flow of money) creates a magnetic field (economic analog = active human population) which, if the current (money flow) begins to diminish, collapse (war) to maintain the current (flow of money - energy).

Other large alternatives to war as economic inductors or economic flywheels are an open-ended social welfare program, or an enormous (but fruitful) open-ended space program. The problem with stabilizing the economic system is that there is too much demand on account of (1) too much greed and (2) too much population. This creates excessive economic inductance which can only be balanced with economic capacitance (true resources or value - e.g., in goods or services).

The social welfare program is nothing more than an open-ended credit balance system which creates a false capital industry to give nonproductive people a roof over their heads and food in their stomachs. This can be useful, however, because the recipients become state property in return for the "gift," a standing army for the elite. For he who pays the piper picks the tune. Those who get hooked on the economic drug must go to the elite for a fix. In this, the method of introducing large amounts of stabilizing capacitance is by borrowing on the future "credit" of the world. This is a fourth law of motion - onset, and consists of performing an action and leaving the system before the reflected reaction returns to the point of action - a delayed reaction.

The means of surviving the reaction is by changing the system before the reaction can return. By this means, politicians become more popular in their own time and the public

pays later. In fact, the measure of such a politician is the delay time. The same thing is achieved by a government by printing money beyond the limit of the gross national product, and economic process called inflation. This puts a large quantity of money into the hands of the public and maintains a balance against their greed, creates a false self-confidence in them and, for awhile, stays the wolf from the door. They must eventually resort to war to balance the account, because war ultimately is merely the act of destroying the creditor, and the politicians are the publicly hired hit men that justify the act to keep the responsibility and blood off the public conscience. (See section on consent factors and social-economic structuring.)

If the people really cared about their fellow man, they would control their appetites (greed, procreation, etc.) so that they would not have to operate on a credit or welfare social system which steals from the worker to satisfy the bum. Since most of the general public will not exercise restraint, there are only two alternatives to reduce the economic inductance of the system.

1. Let the populace bludgeon each other to death in war, which will only result in a total destruction of the living earth.

2. Take control of the world using economic "silent weapons" in a form of "quiet warfare" and reduce the economic inductance of the world to a safe level by a process of benevolent slavery and genocide.

The latter option has been taken as the obviously better option. At this point it should be crystal clear to the reader why absolute secrecy about the silent weapons is necessary. The general public refuses to improve its own mentality and its faith in its fellow man. It has become a herd of proliferating barbarians, and, so to speak, a blight upon the face of the earth.

They do not care enough about economic science to learn why they have not been able to avoid war despite religious morality, and their religious or self-gratifying refusal to deal with earthly problems renders the solution of the earthly problem unreachable to them. It is left to those few who are truly willing to think and survive as the fittest to survive, to solve the problem for themselves as the few who really care. Otherwise, exposure of the silent weapon would destroy our only hope of preserving the seed of the future true humanity.

Industry Equivalent Circuits

The industry 'Q' can be given a block symbol as follows:

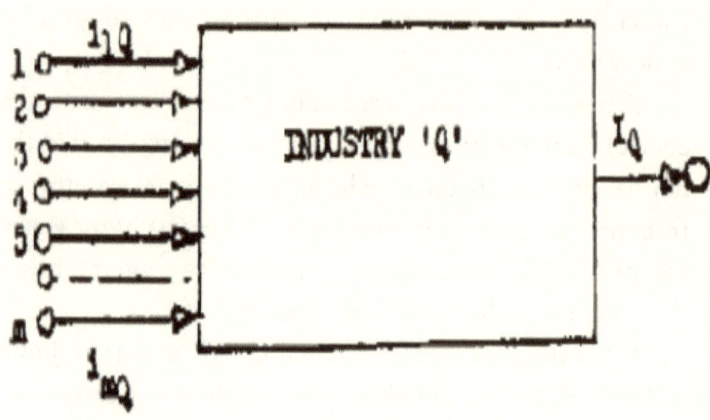

Block Diagram of Industry 'Q".

Terminals #1 through #m are connected directly to the outputs of industries #1 and #m,

respectively.

The equivalent circuit of industry 'Q' is given as follows:

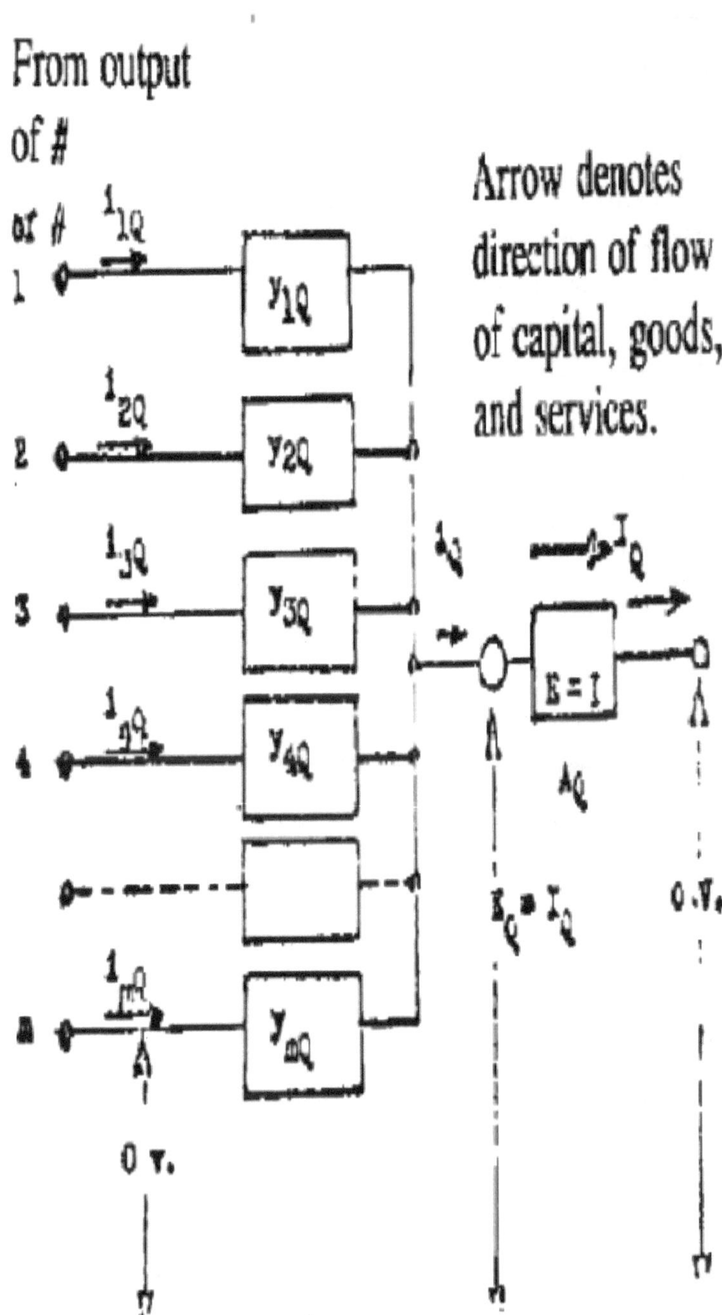
From output
of #
or #
1
2
3
4
a
Arrow denotes
direction of flow
of capital, goods,
and services.
0 v.
0 v.

Characteristics:

All inputs are at zero volts.

A - Amplifier - causes output current IQ to be represented by a voltage EQ. Amplifier

delivers sufficient current at EQ to drive all loads Y10 through YmQ and sink all currents

i1Q through imQ.

The unit trans conductance amplifier AQ is constructed as follows:

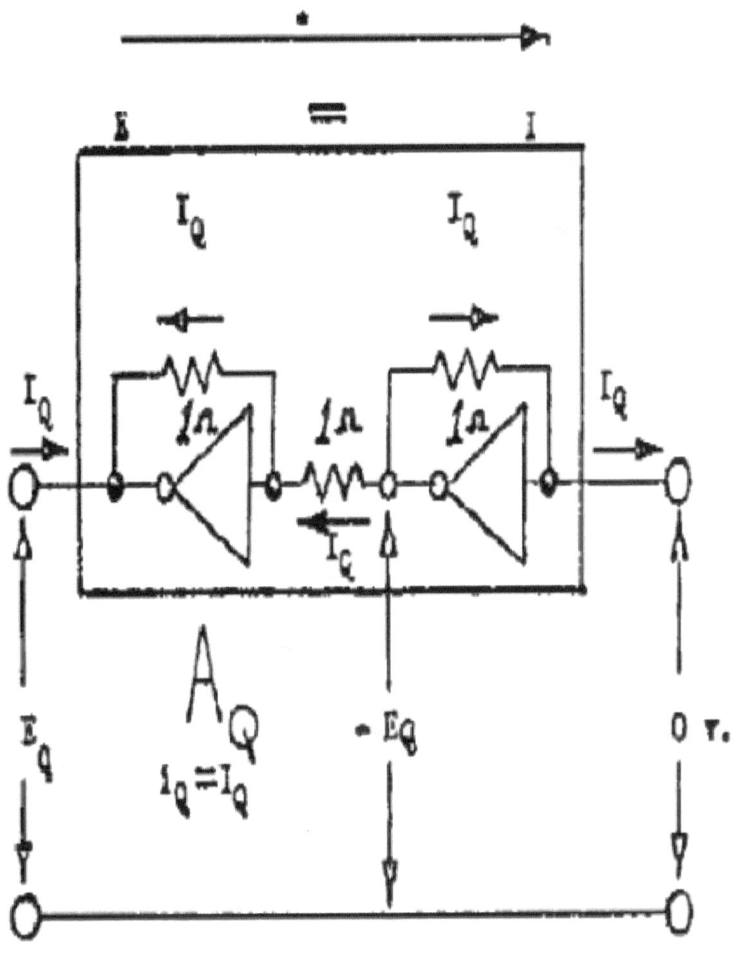

- Arrow denotes the direction of the flow of capital, goods, and services. The total demand is given as EQ, where EQ=IQ.

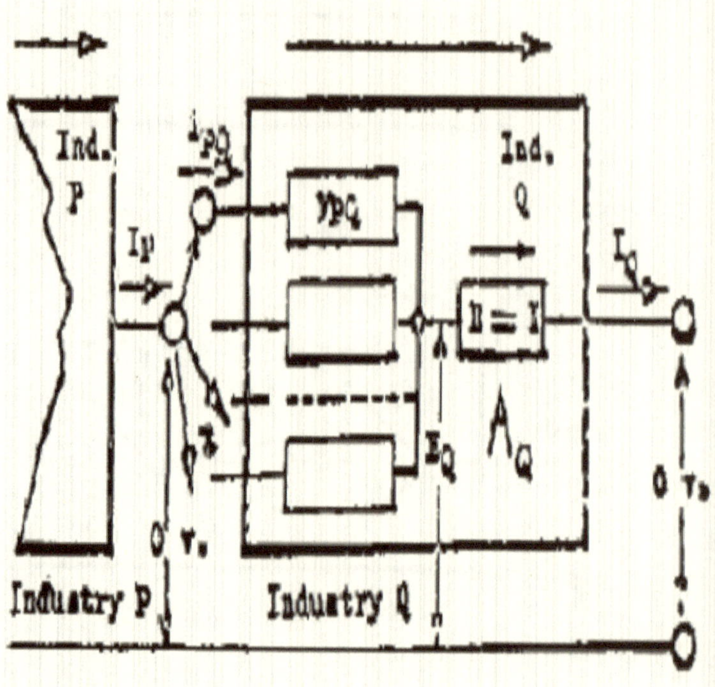

The coupling network YPQ symbolizes the demand which industry Q makes on industry P. the connective admittance YPQ is called the 'technical coefficient' of the industry Q stating the demand of industry Q, called the industry of use, for the output in capital, goods, or services of industry P called the industry of origin.

The flow of commodities from industry P to industry Q is given by iPQ evaluated by the formula:

iPQ = YPQ* EQ.

When the admittance YPQ is a simple conductance, this formula takes on the common appearance of Ohm's Law,

iPQ = gPQ* IQ.

The interconnection of a three-industry system can be diagrammed as follows. The blocks of the industry diagram can be opened up revealing the technical coefficients, and a much simpler format. The equations of flow are given as follows:

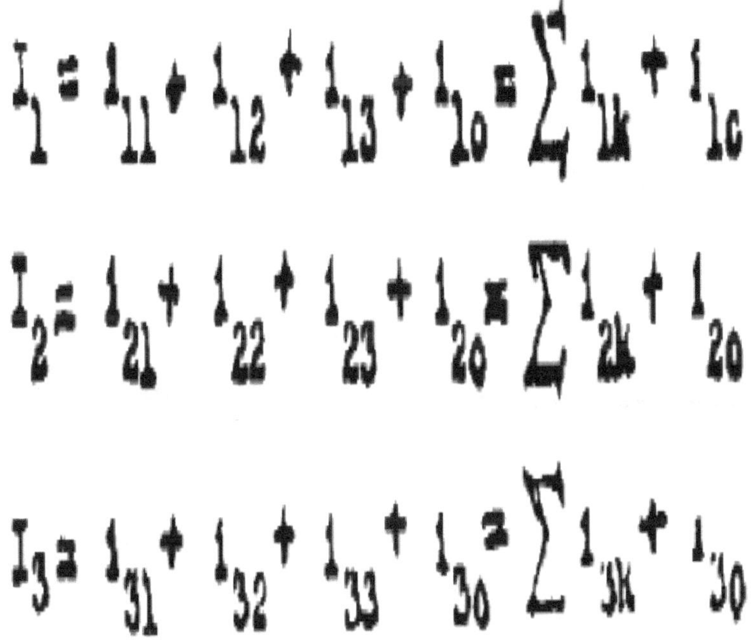

Stages of Schematic Simplification

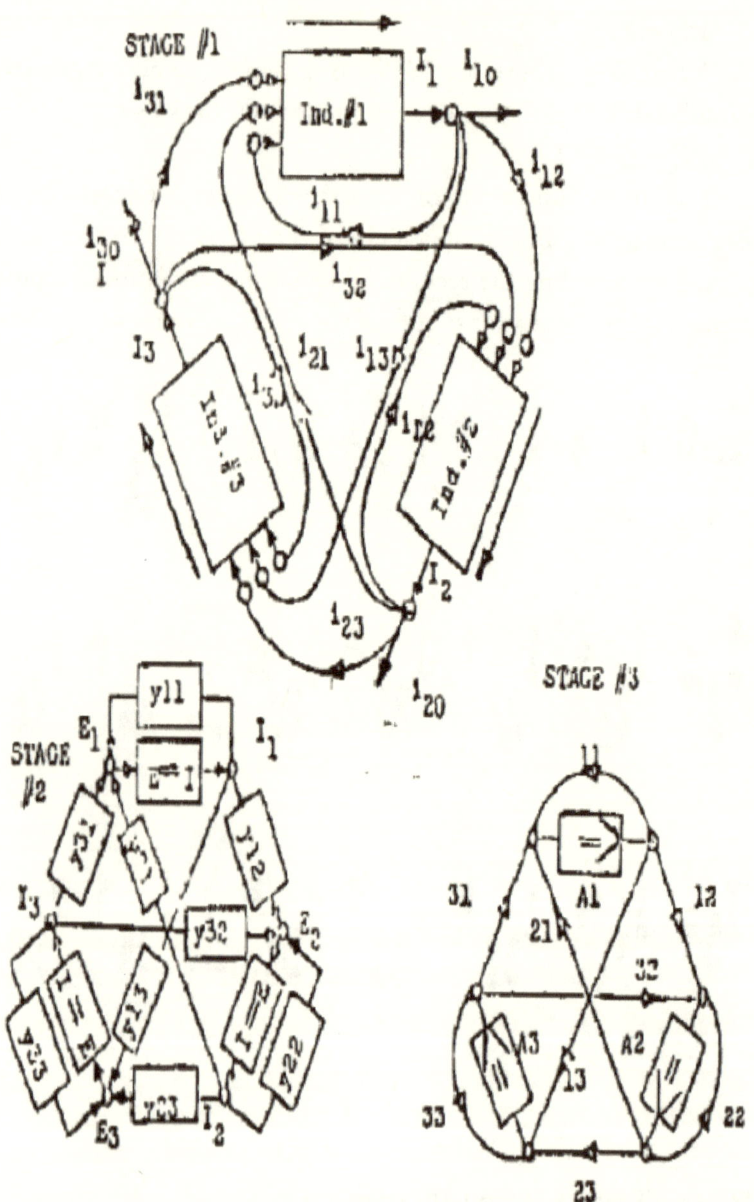
STAGE #1
Ind. #1
I_1
i_{10}
i_{31}
i_{12}
i_{11}
i_{30}
I
i_{32}
I3
i_{21}
i_{13}
Ind. #3
Ind. #2
i_{22}
I_2
i_{23}
i_{20}
STAGE #2
y11
E_1
I_1
y31
y12
I_3
y32
E_2
y13
y33
y21
E_3
I_2
STAGE #3
11
A1
31
12
21
32
A3
A2
13
33
22
23

Generalization

All of this may now be summarized.
Let Ij represent the output of industry j, and

- ijk, the amount of the product of industry j absorbed annually by industry k, and
- ijo, the amount of the same product j made available for 'outside' use. Then

$$I_j = i_{j1} + i_{j2} + i_{j3} + \cdots + i_{jn} + i_{jo}$$

$$= \sum_{k=1}^{k=n} i_{jk} + i_{jo}$$

Substituting the technical co efficiencies, yjk

$$i_{jk} = Y_{jk} I_k$$

$$I_j = \sum_{k=1}^{k=m} i_{jk} + i_{jo} = \sum_{k=1}^{k=m} Y_{jk} I_k + i_{jo}$$

Leontief
Matrix for
$j = 1,2,3,\ldots m$
$$\left\{ I_j - \sum_{k=1}^{k=m} Y_{jk} I_k = i_{jo} \right.$$

Let I_k at the output of industry k be represented by a demand voltage E_k at its amplifier input, i.e., let $E_k = I_k$. Then

$$i_{jk} = Y_{jk} E_k$$

which is the general equation of every admittance in the industry circuit.

Final Bill of Goods

$$\sum_{j=1}^{j=m} i_{j0} = i_{10} + i_{20} + i_{30} + \cdots + i_{m0} \quad \text{is called}$$

is called the final bill of goods or the bill of final demand and is zero when the system can be closed by the evaluation of the technical coefficients of the 'non-productive' industries, government and households. Households may be regarded as a productive industry with labor as its output product.

The Technical Coefficients

The quantities yjk are called the technical coefficients of the industrial system. They are admittances and can consist of any combination of three passive parameters, conductance, capacitance, and inductance. Diodes are used to make the flow unidirectional and point against the flow.

- gjk = economic conductance, absorption coefficient
- yjk = economic capacitance, capital coefficient
- Ljk = economic inductance, human activity coefficient

Types of Admittance

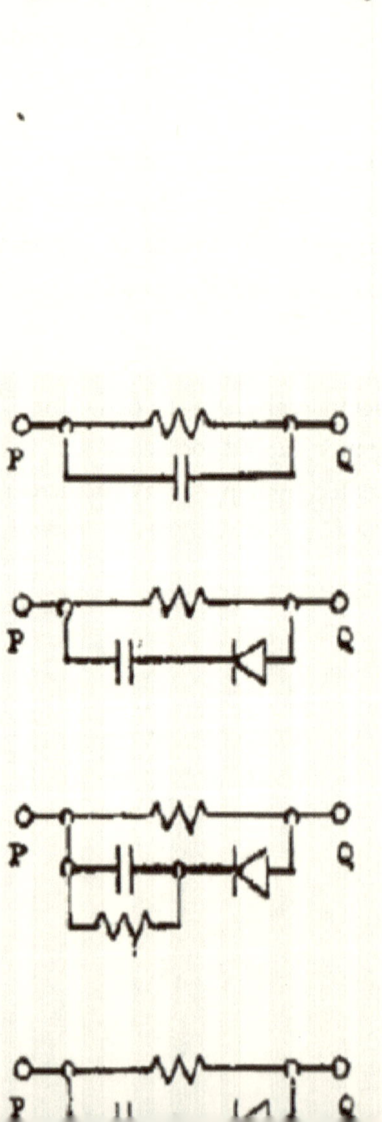

flow of product

storage in industry Q
of capital -- in the
form of inventory of
materials, stock of
equipment, work in
progress, intermediate
products,etc.. This

stock fully reversible
meaning that it <u>can</u> be
sold or exchanged for
other materials.

-flow and stock control,
stock is fully rever-
sible, e.g., <u>con</u> be
sold or exchanged for
other materials.
-flow, but stock not
reversible,
stock does not need
maintenance.

-here the stock is not
reversible, and it is
subject to depreciation.
-can also represent
capital tied up in
buildings which cannot
be sold and are aging.
-- here we have partially
reversible stock which

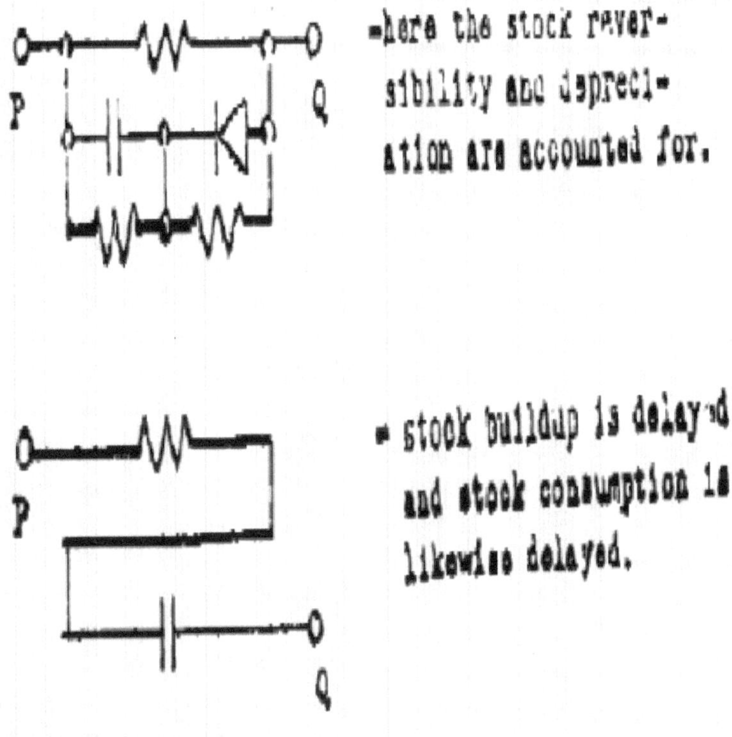

The Household Industry

The industries of finance (banking), manufacturing, and government, real counterparts of the pure industries of capital, goods, and services, are easily defined because they are generally logically structured. Because of this their processes can be described mathematically, and their technical coefficients can be easily deduced. This, however, is not the case with the service industry known as the household industry.

Household Models

When the industry flow diagram is represented by a 2-block system
of households on the right and all other industries on the left, the
following results:

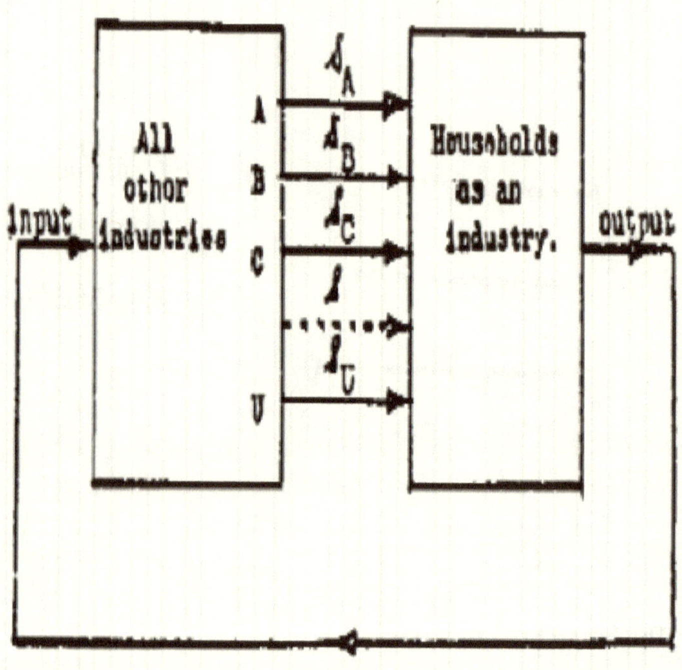

The arrows from left to right labeled A, B, C, etc., denote flow
of economic value from the industries in the left hand block to
the industry in the right hand block called 'households'. These may
be thought of as the monthly consumer flows of the following

commodities. A - alcoholic beverages, B - beef, C - coffee, ... , U - unknown, etc. . . The problem which a theoretical economist faces is that the consumer preferences of any household is not easily predictable and the technical coefficients of any one household tend to be a nonlinear, very complex, and variable function of income, prices, etc.

Computer information derived from the use of the universal product code in conjunction with credit-card purchase as an individual household identifier could change this state of affairs, but the U.P.C. method is not yet available on a national or even a significant regional scale. To compensate for this data deficiency, an alternate indirect approach of analysis has been adopted known as economic shock testing. This method, widely used in the aircraft manufacturing industry, develops an aggregate statistical sort of data. Applied to economics, this means that all of the households in one region or in the whole nation are studied as a group or class rather than individually, and the mass behavior rather than the individual behavior is used to discover useful estimates of the technical coefficients governing the economic structure of the hypothetical single household industry...Notice in the industry flow diagram that the values for the flows A, B, C, etc. are accessible to measurement in terms of selling prices and total sales of commodities. One method of evaluating the technical coefficients of the household industry depends upon shocking the prices of a commodity and noting the changes in the sales of all of the commodities.

Economic Shock Testing

In recent times, the application of Operations Research to the study of the public economy has been obvious for anyone who understands the principles of shock testing. In the shock testing of an aircraft airframe, the recoil impulse of firing a gun mounted on that airframe causes shock waves in that structure which tell aviation engineers the conditions under which some parts of the airplane or the whole airplane or its wings will start to vibrate or flutter like a guitar string, a flute reed, or a tuning fork, and disintegrate or fall apart in flight.

Economic engineers achieve the same result in studying the behavior of the economy and the consumer public by carefully selecting a staple commodity such as beef, coffee, gasoline, or sugar, and then causing a sudden change or shock in its price or availability, thus kicking everybody's budget and buying habits out of shape. They then observe the shock waves which result by monitoring the changes in advertising, prices, and sales of that and other commodities. The objective of such studies is to acquire the know-how to set the public economy into a predictable state of motion or change, even a controlled self-destructive state of motion which will convince the public that certain "expert" people should take control of the money system and reestablish security (rather than liberty and justice) for all. When the subject citizens are rendered unable to control their financial affairs, they, of course, become totally enslaved, a source of cheap labor. Not only the prices of commodities, but also the availability of labor can be used as the means of shock testing. Labor strikes deliver excellent tests shocks to an economy, especially in the critical service areas of trucking (transportation), communication, public utilities (energy, water, garbage collection), etc. By shock testing, it is found that there is a direct relationship between the availability of money flowing in an economy and the real psychological outlook and response of masses of people dependent upon that availability.

For example, there is a measurable quantitative relationship between the price of gasoline and the probability that a person would experience a headache, feel a need to watch a violent movie, smoke a cigarette, or go to a tavern for a mug of beer. It is most interesting that, by observing and measuring the economic models by which the public tries to run from their problems and escape from reality, and by applying the mathematical theory of Operations Research, it is possible to program computers to predict the most probable combination of created events (shocks) which will bring about complete control and subjugation of the public through a subversion of the public economy (by shaking the plum tree).

Introduction to the Theory of Economic Shock Testing

Let the prices and total sales of commodities be given and symbolized as follows:

Commodities	Price Function	Total Sales
alcoholic beverages	A	𝒮A
beef	B	𝒮B
coffee	C	𝒮C
gasoline	G	𝒮G
sugar	S	𝒮S
tobacco	T	𝒮T
unknown balance	U	𝒮U

Let us assume a simple economic model in which the total number of important (staple) commodities are represented as beef, gasoline, and an aggregate of all other staple commodities which we will call the hypothetical miscellaneous staple commodity 'M' (e.g., M is an aggregate of C, S, T, U, etc.).

Example of Shock Testing

Assume that the total sales, P, of petroleum products can be described by the linear function of the quantities B, G, and M, which are functions of the prices of those respective commodities.

$$P = aPG\, B + aPG\, G + aPM\, M$$

Then where B, G, and M are functions of the prices of beef, gasoline, and miscellaneous, respectively, and aPB, aPG, and aPM are constant coefficients defining the amount by which each of the functions B, G, and M affect the sales, P, of petroleum products. We are assuming that B, G, and M are variables independent of each other. If the availability or price of gasoline is suddenly changed, then G must be replaced by G + G. This causes a change in the petroleum sales from P to P + P. Also, we will assume that B and M remain constant when G changes to G + G.

(P + P) = aPB B + aPG (G + G) + aPMM.

Expanding upon this expression, we get

P + P = aPB B + aPG G + aPG G + aPM M

and subtracting the original value of P we get for the change in P

Change in P = P = aPG G

Dividing by G we get

aPG = P / G .

This is a rate of change in P due only to an isolated change in G, G. In general, ajk is the partial rate of change in the sales effect j due to a change in the causal price function of commodity k. If the interval of time were infinitesimal, this expression would be reduced to the definition of the total differential of a function, P.

For if $a_{jk} = \dfrac{\partial j}{\partial k}$, and if $P = a_{PB}B + a_{PC}C + a_{PM}M$

and B, C, and M are independent variables, then

$$a_{PB} = \frac{\partial P}{\partial B} \quad \text{and}$$

$$dP = \frac{\partial P}{\partial B} dB + \frac{\partial P}{\partial C} dC + \frac{\partial P}{\partial M} dM$$

Integrating, we get

$$P = \int \frac{\partial P}{\partial B} dB + \int \frac{\partial P}{\partial C} dC + \int \frac{\partial P}{\partial M} dM.$$

If the a_{jk} are constant coefficients, then the rates, $\partial j / \partial k$, are constant also and can be taken outside of the integrals. Therefore,

$$P = \frac{\partial P}{\partial B} \int dB + \frac{\partial P}{\partial C} \int dC + \frac{\partial P}{\partial M} \int dM \quad \text{or}$$

$$\boxed{P = \frac{\partial P}{\partial B} B + \frac{\partial P}{\partial C} C + \frac{\partial P}{\partial M} M + K.}$$

When the price of gasoline is shocked, all the coefficients with round G (2G) in the

denominator is evaluated at the same time. If B, G, and M were independent, and

sufficient for description of the economy, then three shock tests would be necessary to

evaluate the system.

There are other factors which may be represented the same way.

For example, the tendency of a docile sub-nation to withdraw under economic pressure

may be given by

$$\phi = \frac{\partial \phi}{\partial G} G + \frac{\partial \phi}{\partial W_P} W_P + \cdots$$

where G is the price of gasoline, WP is the dollars spent per unit time (referenced to say

1939) for war production during 'peace' time, etc. These quantities are presented to a

computer in matrix format as full

AND

$$
\begin{vmatrix}
\dfrac{\partial P}{\partial G} & \dfrac{\partial P}{\partial B} & \cdots & \dfrac{\partial P}{\partial U} \\[2ex]
\dfrac{\partial F}{\partial G} & \dfrac{\partial F}{\partial B} & \cdots & \dfrac{\partial F}{\partial U} \\[2ex]
\cdots & \cdots & \cdots & \cdots \\[1ex]
\dfrac{\partial T}{\partial G} & \dfrac{\partial T}{\partial B} & \cdots & \dfrac{\partial T}{\partial U} \\[2ex]
\dfrac{\partial \phi}{\partial G} & \dfrac{\partial \phi}{\partial B} & \cdots & \dfrac{\partial \phi}{\partial U}
\end{vmatrix}
\begin{vmatrix} G \\[2ex] B \\[2ex] \cdot \\[2ex] \cdot \\[2ex] U \end{vmatrix}
=
\begin{vmatrix}
P - K_P \\[2ex]
F - K_F \\[2ex]
\cdots \\[2ex]
T - K_T \\[2ex]
\phi - K_\phi
\end{vmatrix}
$$

$$
\text{or} \quad \left[a_{jk} \right] \left[X_k \right] = \left[Y_j \right]
\qquad\qquad X_k \qquad Y_j
$$

$$
\text{where the } a_{jk} \text{ are defined by} \quad a_{jk} = \frac{\partial X_j}{\partial X_k} \; .
$$

$$
X1 = G \quad Y1 = P - KP
$$
$$
X2 = B \quad Y2 = F - KF
$$
$$
X3 = \text{etc.} \quad Y3 = \text{etc.}
$$

Finally, inverting this matrix, i.e., solving for the Xk terms of the Yj, we get, say, [bkj] [Yj] = [Xk]

This is the result into which we substitute to get that set of conditions of prices of commodities, bad news on TV, etc., which will deliver a collapse of public morale ripe for take over. Once the economic price and sales coefficients ajk and bkj are determined, they may be translated into the technical supply and demand coefficients gjk, Cjk, and 1/Ljk. Shock testing of a given commodity is then repeated to get the time rate of change of these technical coefficients.

Introduction to Economic Amplifiers

Economic amplifiers are the active components of economic engineering. The basic characteristic of any amplifier (mechanical, electrical, or economic) is that it receives an input control signal and delivers energy from an independent energy source to a specified output terminal in a predictable relationship to that input control signal. The simplest form of an economic amplifier is a device called advertising. If a person is spoken to by a T.V. advertiser as if he were a twelve-year-old, then, due to suggestibility, he will, with a certain probability, respond or react to that suggestion with the uncritical response of a twelve-year-old and will reach into his economic reservoir and deliver its energy to buy that product on impulse when he passes it in the store. An economic amplifier may have several inputs and output. Its response might be instantaneous or delayed. Its circuit symbol might be a rotary switch if its options are exclusive, qualitative, "go" or "no-go", or it might have its parametric input/output relationships specified by a matrix with internal energy sources represented.

Whatever its form might be, its purpose is to govern the flow of energy from a source to an output sink in direct relationship to an input control signal. For this reason, it is called an active circuit element or component. Economic Amplifiers fall into classes called strategies, and, in comparison with electronic amplifiers, the specific internal functions of an economic amplifier are called logistical instead of electrical. Therefore, economic amplifiers not only deliver power gain but also, in effect, are used to cause changes in the economic circuitry.

In the design of an economic amplifier, we must have some idea of at least five functions,

which are:

(1) the available input signals,
(2) the desired output-control objectives,
(3) the strategic objective,
(4) the available economic power sources,
(5) the logistical options.

The process of defining and evaluating these factors and incorporating the economic amplifier into an economic system has been popularly called game theory.

The design of an economic amplifier begins with a specification of the power level of the output, which can range from personal to national. The second condition is accuracy of response, i.e., how accurately the output action is a function of the input commands. High gain combined with strong feedback helps to deliver the required precision. Most of the error will be in the input data signal. Personal input data tends to be specified, while national input data tends to be statistical.

Short List of Inputs

Questions to be answered:

- what
- where
- why
- when
- how
- who

General sources of information:

- telephone taps
- analysis of garbage

- surveillance
- behavior of children in school

Standard of living by:

- food
- shelter
- clothing
- transportation

Social contacts:

- telephone - itemized record of calls
- family - marriage certificates, birth certificates, etc.
- friends, associates, etc.
- memberships in organizations
- political affiliation

The Personal Paper Trail

Personal buying habits, i.e., personal consumer preferences:

- checking accounts
- credit-card purchases
- "tagged" credit-card purchases - the credit-card purchase of products bearing the

U.P.C. (Universal Product Code)

Assets:

- checking accounts
- savings accounts

- real estate
- business
- automobile, etc.
- safety deposit at bank
- stock market

Liabilities:

- creditors
- enemies (see - legal)
- loans

Government sources (ploys)*:

- Welfare
- Social Security
- U.S.D.A. surplus food
- doles
- grants
- subsidies
- Principle of this ploy—the citizen will almost always make the collection of information easy if he can operate on the "free sandwich principle" of "eat now and pay later."

Government sources (via intimidation):

- Internal Revenue Service
- OSHA
- Census
- etc.

Other government sources—surveillance of U.S. mail.
Habit Patterns—Programming

Strengths and weaknesses:

- activities (sports, hobbies, etc.)
- see "legal" (fear, anger, etc.—crime record)
- hospital records (drug sensitivities, reaction to pain, etc.)
- psychiatric records (fears, angers, disgusts, adaptability, reactions to stimuli, violence, suggestibility or hypnosis, pain, pleasure, love, and sex)

Methods of coping—of adaptability—behavior:

- consumption of alcohol
- consumption of drugs
- entertainment
- religious factors influencing behavior
- other methods of escaping from reality

Payment modus operandi (MO)—pay on time, etc.:

- payment of telephone bills
- energy purchases
- water purchases
- repayment of loans
- house payments
- automobile payments
- payments on credit cards

Political sensitivity:

- beliefs
- contacts
- position
- strengths/weaknesses

- projects/activities

Legal inputs—behavioral control (Excuses for investigation, search, arrest, or
employment of force to modify behavior)

- court records
- police records—NCIC
- driving record
- reports made to police
- insurance information
- anti-establishment acquaintances

National Input Information

Business sources (via I.R.S., etc.):

- prices of commodities
- sales
- investments in

o stocks/inventory
o production tools and machinery
o buildings and improvements
o the stock market

Banks and credit bureaus:

- credit information
- payment information

Miscellaneous sources:

- polls and surveys
- publications
- telephone records
- energy and utility purchase short List of Outputs

Short List of Outputs

Outputs—create controlled situations—manipulation of the economy, hence society—

control by control of compensation and income.

Sequence:

1. allocates opportunities
2. destroys opportunities
3. controls the economic environment
4. controls the availability of raw materials
5. controls capital.
6. controls bank rates
7. controls the inflation of the currency
8. controls the possession of property
9. controls industrial capacity
10. controls manufacturing
11. controls the availability of goods (commodities).
12. controls the prices of commodities.
13. controls services, the labor force, etc.
14. controls payments to government officials.
15. controls the legal functions.
16. controls the personal data files—uncorrectable by the party slandered.
17. controls advertising.
18. controls media contact.
19. controls material available for T.V. viewing
20. disengages attention from real issues.
21. engages emotions.

22. creates disorder, chaos, and insanity.

23. controls design of more probing tax forms.

24. controls surveillance.

25. controls the storage of information.

26. develops psychological analyses and profiles of individuals.

27. controls legal functions [repeat of 15]

28. controls sociological factors.

29. controls health options.

30. preys on weakness.

31. cripple strengths.

32. leaches wealth and substance.

Table of Strategies

Do This	To Get This
Keep the public ignorant	Less public organization
Maintain access to control point for feedback	Required reaction to outputs (prices, sales)
Create preoccupation	Lower defense
Attack the family unit	Control of the education of the young
Give less cash and more credit and doles	More self-indulgence and more data
Attack the privacy of the church	Destroy faith in this sort of government
Social conformity	Computer programming simplicity
Minimize the tax protest	Maximum economic data, minimum enforcement problems

Stabilize the consent	Simplicity coefficients
Tighten control of variables	Simpler computer input data - greater predictability
Establish boundary conditions	Problem simplicity/solutions of differential and difference equations
Proper timing	Less data shift and blurring
Maximize control	Minimum resistance to control
Collapse of currency	Destroy the faith of the American people in each other

Diversion, the Primary Strategy

Experience has presented that the simplest method of securing a silent weapon and gaining control of the public is to keep the public undisciplined and ignorant of the basic system principles on the one hand, while keeping them confused, disorganized, and distracted with matters of no real importance on the other hand.

This is achieved by:

- disengaging their minds; sabotaging their mental activities; providing a low-quality program of public education in mathematics, logic, systems design and economics; and discouraging technical creativity.
- engaging their emotions, increasing their self-indulgence and their indulgence in emotional and physical activities, by: o unrelenting emotional confrontations and attacks (mental and emotional rape) by way of constant barrage of sex, violence, and wars in the media - especially the T.V. and the newspapers. o giving them what they desire - in excess - "junk food for thought" - and depriving them of what they really need.
- rewriting history and law and subjecting the public to the deviant creation, thus being able to shift their thinking from personal needs to highly fabricated outside priorities.

These preclude their interest in and discovery of the silent weapons of social automation technology. The general rule is that there is a profit in confusion, the more confusion, the more profit. Therefore, the best approach is to create problems and then offer solutions.

Diversion Summary

Media: Keep the adult public attention diverted away from the real social issues and captivated by matters of no real importance.

Schools: Keep the young public ignorant of real mathematics, real economics, real law, and real history.

Entertainment: Keep the public entertainment below a sixth-grade level.

Work: Keep the public busy, busy, busy, with no time to think; back on the farm with the other animal Consent, the Primary Victory.

A silent weapon system operates upon data obtained from a docile public by legal (but not always lawful) force. Much information is made available to silent weapon systems programmers through the Internal Revenue Service. (See Studies in the Structure of the American Economy for an I.R.S. source list.)

This information consists of the enforced delivery of well-organized data contained in federal and state tax forms, collected, assembled, and submitted by slave labor provided by taxpayers and employers. Furthermore, the number of such forms submitted to the I.R.S. is a useful indicator of public consent, an important factor in strategic decision making. Other data sources are given in the Short List of Inputs.

Consent Coefficients - numerical feedback indicating victory status. Psychological basis:

When the government can collect tax and seize private property without just compensation, it is an indication that the public is ripe for surrender and is consenting to enslavement and legal encroachment. A good and easily quantified indicator of harvest time is the number of public citizens who pay income tax despite an obvious lack of reciprocal or honest service from the government.

Amplification Energy Sources

The next step in the process of designing an economic amplifier is discovering the energy sources. The energy sources which support any primitive economic system are, of course, a supply of raw materials, and the consent of the people to labor and consequently assume a certain rank, position, level, or class in the social structure, i.e., to provide labor at various levels in the pecking order. Each class, in guaranteeing its own level of income, controls the class immediately below it, hence preserves the class structure. This provides stability and security, but also government from the top. As time goes on and communication and education improve, the lower-class elements of the social labor structure become knowledgeable and envious of the good things that the upper-class members have. They also begin to attain a knowledge of energy systems and the ability to enforce their rise through the class structure. This threatens the sovereignty of the elite. If this rise of the lower classes can be postponed long enough, the elite can achieve energy dominance, and labor by consent no longer will hold a position of an essential energy source. Until such energy dominance is absolutely established, the consent of people to labor and let others handle their affairs must be taken into consideration, since failure to do so could cause the people to interfere in the final transfer of energy sources to the control of the elite. It is essential to recognize that at this time, public consent is still an essential key to the release of energy in the process of economic amplification. Therefore, consent as an energy release mechanism will now be considered.

Logistics

The successful application of a strategy requires a careful study of inputs, outputs, the strategy connecting the inputs and the outputs, and the available energy sources to fuel the strategy. This study is called logistics. A logistical problem is studied at the elementary level first, and then levels of greater complexity are studied as a synthesis of elementary factors. This means that a given system is analyzed, i.e., broken down into its subsystems, and these in turn are analyzed, until by this process, one arrives at the logistical "atom," the individual. This is where the process of synthesis properly begins, at the time of birth of the individual.

The Artificial Womb

From the time a person leaves its mother's womb, its every effort is directed towards building, maintaining, and withdrawing into artificial wombs, various sorts of substitute protective devices or shells. The objective of these artificial wombs is to provide a stable environment for both stable and unstable activity; to provide a shelter for the evolutionary processes of growth and maturity - i.e., survival; to provide security for freedom and to provide defensive protection for offensive activity.

This is equally true of both the general public and the elite. However, there is a definite difference in the way each of these classes go about the solution of problems.

The Political Structure of a Nation – Dependency

The primary reason why the individual citizens of a country create a political structure is a subconscious wish or desire to perpetuate their own dependency relationship of childhood. Simply put, they want a human god to eliminate all risk from their life, pat them on the head, kiss their bruises, put a chicken on every dinner table, clothe their bodies, tuck them into bed at night, and tell them that everything will be alright when they wake up in the morning. This public demand is incredible, so the human god, the politician, meets incredibility with incredibility by promising the world and delivering nothing. So, who is the bigger liar? the public? or the "godfather"? This public behavior is surrender born of fear, laziness, and expediency. It is the basis of the welfare state as a strategic weapon, useful against a disgusting public.

Action/Offense

Most people want to be able to subdue and/or kill other human beings which disturb their daily lives, but they do not want to have to cope with the moral and religious issues which such an overt act on their part might raise. Therefore, they assign the dirty work to others (including their own children) to keep the blood off their hands. They rave about the humane treatment of animals and then sit down to a delicious hamburger from a whitewashed slaughterhouse down the street and out of sight. But even more hypocritical, they pay taxes to finance a professional association of hit men collectively called politicians, and then complain about corruption in government. Responsibility, again, most people want to be free to do the things (to explore, etc.) but they are afraid to

fail. The fear of failure is manifested in irresponsibility, and especially in delegating those personal responsibilities to others where success is uncertain or carries possible or created

liabilities (law), which the person is not prepared to accept. They want authority (root word - "author"), but they will not accept responsibility or liability. So they hire politicians to face reality for them.

Summary

The people hire the politicians so that the people can:

(1) obtain security without managing it.
(2) obtain action without thinking about it.

(3) inflict theft, injury, and death upon others without having to contemplate either life or death.

(4) avoid responsibility for their own intentions.

(5) obtain the benefits of reality and science without exerting themselves in the discipline of facing or learning either of these things.

They give the politicians the power to create and manage a war machine:

(1) provide for the survival of the nation/womb.

(2) prevent encroachment of anything upon the nation/womb.

(3) destroy the enemy who threatens the nation/womb.

(4) destroy those citizens of their own country who do not conform for the sake of stability of the nation/womb.

Politicians hold many quasi-military jobs, the lowest being the police who are soldiers, the attorneys and C.P.A.s next who are spies and saboteurs (licensed), and the judges who shout orders and run the closed union military shop for whatever the market will bear. The generals are industrialists. The "presidential" level of commander-in-chief is shared by the international bankers. The people know that they have created this farce and financed it with their own taxes (consent), but they would rather knuckle under than be the hypocrite. Thus, a nation becomes divided into two very distinct parts, a docile sub-nation [great silent majority] and a political sub-nation. The political sub-nation remains attached to

the docile sub-nation, tolerates it, and leaches its substance until it grows strong enough to detach itself and then devour its parent.

System Analysis

In order to make meaningful computerized economic decisions about war, the primary economic flywheel, it is necessary to assign concrete logistical values to each element of the war structure -

personnel and material alike. This process begins with a clear and candid description of the subsystems of such a structure.

The Draft (As military service)

Few efforts of human behavior modification are more remarkable or more effective than that of the socio-military institution known as the draft. A primary purpose of a draft or other such institution is to instill, by intimidation, in the young males of a society the uncritical conviction that the government is omnipotent. He is soon taught that a prayer is slow to reverse what a bullet can do in an instant. Thus, a man trained in a religious environment for eighteen years of his life can, by this instrument of the government, be broken down, be purged of his fantasies and delusions in a matter of mere months. Once that conviction is instilled, all else becomes easy to instill. Even more interesting is the process by which a young man's parents, who purportedly love him, can be induced to send him off to war to his death. Although the scope of this work will not allow this matter to be expanded in full detail, nevertheless, a coarse overview will be possible and can serve to reveal those factors which must be included in some numerical form in a computer analysis of social and war systems. We begin with a tentative definition of the draft. The draft (selective service, etc.) is an institution of compulsory collective sacrifice and slavery, devised by the middle-aged and elderly for the purpose of pressing the young into doing the public dirty work. It further serves to make the youth as guilty as the elders, thus making criticism of the elders by the youth less likely (Generational Stabilizer). It is marketed and sold to the public under the label of "patriotic = national" service. Once a candid economic definition of the draft is achieved, that definition is used to outline the boundaries of a structure called a Human Value System, which in turn is translated into the terms of game theory. The value of such a slave laborer is given in a Table of Human Values, a table broken down into categories by intellect, experience, post-service job demand, etc. Some of these categories are ordinary and can be tentatively evaluated in terms of the value of certain jobs for which a known fee exists. Some jobs are harder to value because they are unique to the demands of social subversion, for an extreme example: the value of a mother's instruction to her daughter, causing that daughter to put certain behavioral demands upon a future husband ten or fifteen years; hence, thus, by suppressing his resistance to a perversion of a government, making it easier for a banking cartel to buy the State of New York in, say, twenty years. Such a problem leans heavily upon the observations and data of wartime espionage and many types of psychological testing. But crude mathematical models (algorithms, etc.) can be devised, if not to predict, at least to predeterminate these events with maximum certainty. What does not exist by natural cooperation is thus enhanced by calculated compulsion. Human beings are machines, levers which may be grasped and turned, and there is little real difference between automating a society and automating a shoe factory. These derived values are variable. (It is necessary to use a current Table of Human Values for computer analysis.) These values are given in true measure rather than U.S. dollars, since the latter is unstable, being presently inflated beyond the production of national goods and services to give the economy a false kinetic energy ("paper" inductance). The silver value is stable, it being possible to buy the same amount with a gram of silver today as it could be bought in 1920. Human value measured in silver units changes slightly due to changes in production technology.

Enforcement

Factor I

As in every social system approach, stability is achieved only by understanding and accounting for human nature (action/reaction patterns). A failure to do so can be, and usually is, disastrous. As in other human social schemes, one form or another of intimidation (or incentive) is essential to the success of the draft. Physical principles of action and reaction must be applied to both internal and external subsystems.

To secure the draft, individual brainwashing/programming and both the family unit and the peer group must be engaged and brought both under control.

Factor II

Father - The man of the household must be housebroken to ensure that junior will grow up with the right social training and attitudes. The advertising media, etc., are engaged to see to it that father-to-be is pussy-whipped before or by the time he is married. He is taught that he either conforms to the social notch cut out for him or his sex life will be hobbled, and his tender companionship will be zero. He is made to see that women demand security more than logical, principled, or honorable behavior. By the time his son must go to war, father (with jelly for a backbone) will slam a gun into junior's hand before father will risk the censure of his peers or make a hypocrite of himself by crossing the investment he has in his own opinion or self-esteem. Junior will go to war or father will be embarrassed. So, junior will go to war, the true purpose notwithstanding.

Factor III

Mother - The female element of human society is ruled by emotion first and logic second. In the battle between logic and imagination, imagination always wins, fantasy prevails, maternal instinct dominates so that the child comes first, and the future comes second. A woman with a newborn baby is too starry-eyed to see a wealthy man's cannon fodder or a cheap source of slave labor. A woman must, however, be conditioned to accept the transition to "reality" when it comes, or sooner. As the transition becomes more difficult to manage, the family unit must be carefully disintegrated, and state-controlled public education and state operated child-care centers must be become more common and legally enforced to begin the detachment of the child from the mother and father at an earlier age. Inoculation of behavioral drugs [Ritalin] can speed the transition for the child (mandatory). Caution: A woman's impulsive anger can override her fear. An irate woman's power must never be underestimated, and her power over a pussy-whipped husband must likewise never be underestimated. It got women the vote in 1920.

Factor IV – Junior

The emotional pressure for self-preservation during the time of war and the self-serving attitude of the common herd that have an option to avoid the battlefield - if junior can be persuaded to go - is all of the pressure finally necessary to propel Johnny off to war. Their quiet black mailings of him are the threats: "No sacrifice, no friends; no glory, no girlfriends.

Factor V – Sister

And what about junior's sister? She is given all the good things of life

her father, and taught to expect the same from her future husband

regardless of the price.

Factor VI – Cattle

Those who will not use their brains are no better off than those who have no brains, and so this mindless school of jelly-fish, father, mother, son, and daughter, become useful beasts of burden or trainers of the same.

This concludes what is available of this document

i "Studies in the Structure of American Economy" (1953), by Wassily Leontief (Director of the Harvard Economic Research Project), International Science Press Inc., White Plains, New York.

"No people will tamely surrender their Liberties, nor can any be easily subdued, when knowledge is diffused and Virtue is preserved. On the Contrary, when People are universally ignorant, and debauched in their Manners, they will sink under their own weight without the Aid of foreign Invaders."

Samuel Adams

"For we are opposed around the world by a monolithic and ruthless conspiracy that relies on covert means for expanding its sphere of influence--on infiltration instead of invasion, on subversion instead of elections, on intimidation instead of free choice, on guerrillas by night instead of armies by day. It is a system that has conscripted vast human and material resources into the building of a tightly knit, highly efficient machine that combines military, diplomatic, intelligence, economic, scientific and political operations."

John F. Kennedy- SPEECH

"The individual is handicapped, by coming face-to-face, with a conspiracy so monstrous, he cannot believe it exists. The   American mind, simply has not come to a realization of the evil, which has been introduced into our midst . . . It rejects even the assumption that human creatures could espouse a philosophy, which must ultimately destroy all that is good and decent."

FBI Director J. Edgar Hoover, 1956

"A silent weapon system operates upon data obtained from a docile public by legal (but not always lawful) force. When the government is able to collect tax and seize private property without justification, it is an indication that the public is ripe for surrender and is consenting to enslavement and legal encroachment."

"War is therefore the balancing of the system by killing the true creditors (the public, which we have taught to exchange true value for inflated currency) and falling back on whatever is left of the resources of nature and regeneration of those resources."

"Until such energy dominance is absolutely established, the consent of people to labor and let others handle their affairs must be taken into consideration, since failure to do so could cause the people to interfere in the final transfer of energy sources to the control of the elite. It is essential to recognize that at this time, public consent is still an essential key to the release of energy in the process of economic amplification. Therefore, consent as an energy release mechanism will now be considered."

Quotes from "Silent Weapons for Quiet Wars"

If you have read and followed the commentary for this document, you might be curious enough to follow the trail of evidence which threatens the entire establishment with the killing of John F Kennedy just before he was to address Dallas Texas of the enslavement agenda. There has not been a President since Kennedy to even think in terms of divulging the real truth. What we have witnessed these past 67 years since the proclamation of enslavement was introduced are the telltale signs of corporate control look again at page 165 the third to last control point 29 *Health options* is where we sit today

2022 just three more stages are left if we believe the evidence. Another observation to consider, the numbering the 70 years of Daniels vision we have 3 years to its expiry 1954+70 2024 watch August. But be not fearful of such that practice their art of slavery, debauchery, whore mongering, deceit, and lies. There is a time for all things and our choices are part of the stream of water which flows the life in life and in death to life again when we trust God rather than man.

When you sit back and ponder the implications you know the fate of those who are the delusional you get a kind of empathy for them because you know the truth of the matter. God said pray for them that they may have a chance at life everlasting !! This brings us to the patience of the prayer for them their error be known to them, and change is offered spiritually. I will put forward undisputable evidence which puts a higher power above all these feeble attempts to secure control of the populations. But will they believe? It has always been that root question. Some will disregard this books information and that is their choice.

The day will also arrive when choice will be no more and the mystery of the two witnesses will come into view. Why are they here? What is their purpose? What scripture supports their arrival? What will they accomplish?

Chapter 8

Power for the next 200+ years
The Solution

Many have heard me talk about the proper pathway for our civilization to follow. If I say follow, that means the answer is here and evident today as a primary solution to current problems old thinking has caused. Now we have a solution but it's never a solution if never implemented. The days ahead are what will try the fortitude of good people to do good things. Planetaryhealthproject.org is a good thing. How do we know what is good to do and what could be a ploy or failure moving forward? When we take the best reasoning and step into the arena of peer review this either sets a new standard or removes the standard trying to be set. Let's look at the business aspect of this new energy invention.

First, does it work and how do we know it works? Those are fair questions. When we investigate kinetic energy what is it really? It's defined as a moving force sufficient to do work if the configuration is correct do lots of work. What kind of configuration are we talking about? This particular case uses water fresh water and bringing that water to a certain kinetic energy state which continues to work as long as that energy state is maintained. We have found the answer to this problem of Ebb flow in a water course. Using the entire column of water and transferring the energy of that column to do a large amount of work. Now think of it this way if we divulge the master work to say an individual billionaire what do you think would happen? That billionaire would find a way to steal the entire workings at near no cost and use it for his or her gain, sounds logical enough but that is not delivering the intent of this invention to solving the future problem of energy. If you're a billionaire, you either inherited the some or you made it on your own railroading the competition at every turn of events to protect your interests. That's how billionaires maintain their billions at the cost of the masses which consume their industrialist products. This company is different and approaches energy differently by wanting to give the rights of the invention to the public. Here is some of the reality points.

The Worldwide Development Corporation (WDC) was recently formed after 12 years of pain staking relentless effort, disappointments, self-funding, failures, successes, etc. developing an energy invention that can change the world. WDC has put together a World Class project to introduce and implement a new energy invention that has the capability to power the planet 1000 times over without consuming resources after set up, nonpolluting for a fraction of the cost of any energy generated today theoretically forever. It is the crown jewel of energy generation and the missing piece to the sustainability of humans. The project includes all the Engineering, 3D drawings and Prototype that is professionally done. The quality and thoroughness of engineering completed on this project is second to none. The validation process for the power projections for the new energy invention will be an elementary exercise for any engineer. All math calculations are easily understood and have sound engineering disciplines behind them.

This new energy invention can generate an almost unlimited amount of clean firm 24/7 energy distributed throughout the entire planet without feed stocks and a small ecological footprint for a fraction of the cost of any energy generated today. The new energy invention is based on kinetics which means the energy is generated through a body in motion and not heat or thermodynamics that consume resources.

Now what do I mean by almost unlimited energy?

This new energy invention has the capabilities to generate 10,000 or even 20,000 terawatts of firm 24/7 electricity distributed widely around the entire planet. Today the world generates approximately 6.6 terawatts of electricity for the entire planet. This technology is leap years ahead of any energy generation technology that exist today. It's now time to bring the world together for one mission and that is to introduce and implement this technology all across the planet.

We have the opportunity to change the priorities of the human race for the betterment of mankind. Don't let fear of the future or greed stand in our way. Because of the almost unbelievable capabilities of this new energy invention, we can transform from a society that concentrates on generating energy to how do we use energy to improve the lives of billions of people all across this planet. Even with this new energy invention capabilities the path to create a world that is sustainable will not be easy but if we all join together, we can truly live up to the spirit of humanity at its best.

During the past 4,000 years mankind has been able to survive and thrive without any concern of limited natural resources or population growth. But over the last 70 years the world's population has grown from 1.5 billion people to over 7.4 billion people with no signs of slowing down. This has caused a great strain on our limited world resources that is unprecedented in human history. *Without drastic changes our world as we know it today will cease to exist in the not-too-distant future.*

The ultimate goal of WDC is to change our priorities from a fossil fuel (limited resource) consumption throw away society to an alternative energy recycle and sustainable civilization where humanity can survive forever. It is impossible to power our planet by consuming resources regardless the resource. With a population of 7.8 billion and growing we must revert to reason and plant the regenerative power systems of the future today. We must plan and execute logic and properly planned pathways which slow down our population march by limiting family size or other sensible and fair systems to meet future requirements. You will soon learn of a new method of energy generation that is so ingenious, so exciting, so revolutionary that nothing like it has ever been seen before.

The Worldwide Development Corporation's goal is to collect 10% royalties on the energy generated worldwide from this new energy invention. Then 60% of the royalties collected will be redistributed through a non-profit corporation to assist poor countries in implementing this technology and rebuild a proper power structure around the world. Some type of financial vehicle has to be created to assist 3rd World Countries if we want to create a fairer and more balanced world and this is just as good of a place to start as any to accomplish that task.

We believe this will be the largest non-profit corporation in the world and much emphasis and effort will be given to make sure it's operated with the highest ethical standards. Our plan is to create a selection committee made up of regular citizens from around the world that are not political to make the decisions on what new energy invention installations will be funded based on humanitarian needs.

You may ask yourself what's at stake?

The lives of billions of people with no hope will now have a chance to participate in the reality we call humanity.

The mission of the WDC is to deliver a new energy technology that can generate almost unlimited energy without consuming resources for a fraction of the cost of any energy generated today widely distributed throughout our entire planet. We believe this will be the basis for a sustainable planet for humanity theoretically forever. The domino effect of almost unlimited cheap energy distributed across the planet will be the catalyst to a worldwide 21st century civilization where there is a place for everybody and a chance for hope and prosperity.

Mankind's ultimate responsibility is to create a sustainable planet for humanities future because without it everything else is a moot point.

This challenge will determine if the United Nations is an institution that is willing to lead in the future of humanity or just a body of Nations jockeying for power and influence. I've now added GE and/or Siemens to the challenge if they are brave enough to step forward. The Worldwide Development Corporation has created an alternative energy invention that we believe will change everything we think we know about energy generation. The new energy invention is capable of generating 100,000 terawatts of 24/7 electricity distributed throughout the entire planet without consuming resources for a small fraction of the cost of any energy generated today. Today our planet generates approximately 6.6 terawatts of 24/7 firm electricity. Even though we would never contemplate or even imagine generating 100,000 terawatts of 24/7 electricity worldwide, I thought it was important to point out the extreme capabilities of the new energy invention.

A Sustainable World

Resource management is the only way we can create a sustainable planet for mankind. There are hundreds of things we can do to create a healthier and more prosperous world long term. Two things that are absolutely necessary to have true sustainability are an unlimited energy supply without feed stocks and a 360 recyclable program worldwide that recycles all resources for reuse. The key word is "ALL" resources must be recycled for reuse. Both are critical in order to preserve and sustain the limited resources necessary to support a modern society made up of billions of people long term. The definition of sustainability is the ability to continue to do an action theoretically forever. Either something is sustainable or it's not because there's no in between. new energy invention which is the one missing piece to humanities long term survival that has eluded mankind up till now. This new energy invention will give humanity the opportunity for energy-starved nations to feed their populations, provide them with clean water, greatly reduced air pollution and provide the energy needed to support exploding human populations that are today at 7.5 billion people and projected to be 10 billion people by the year 2050.

The only other missing piece to the sustainability of the human race is recycling all our resources. The only thing required to solve this final missing piece to mankind's sustainability is the *will of the people*. It will be up to the United Nations to lead this endeavor because without every country on the planet participating it will all be in vain.

I think it's important to give an example how a worldwide recycling program could be implemented around the world. At the time of manufacturing products, a recycle fee could be collected that would be distributed to the country where the products are sold to pay for recycling. We need to begin to think of recycling as just part of the cost of a product. Again, I realize the initial startup of this program would require special assistance for poorer countries, etc. But we all should remember the welfare and future of humanity is at stake.

We must begin to think and act larger than ourselves.

We have choices – Dinosaurs made it for almost an hour on the Earth time clock – Humans have made it a little over a minute – *I think we can do better*

Common sense tells you 7.5 billion people and growing consuming LIMITED RESOURCES at an unprecedented rate will not end well for humanity – Remember the world had a population of 1.5 billion people just 70 years ago (1950). You don't need to be a genius to understand humanity is on a path to self-destruction fighting over the last remaining resources – Just try to imagine a world with 10 billion people and a lack of energy and no plastics – *You can't because that world couldn't exist.*

Before we can discuss the solution and benefits to powering the planet without fossil fuels, we must establish there is a problem:

98% of our transportation fuels come from oil. 30% of the oil consumed in U.S. is to make products. When we deplete all the oil reserves we not only lose our transportation fuels, we lose our ability to create a modern society and without a modern society the world would collapse. Try to imagine a world without plastics. If you look closely at the picture above over 7,000 products that make up a modern society is made from oil. There is no easy replacement for oil as a feed stock to make plastics. That's why the importance of preserving oil should be one of the most important priorities in the world if we care about the future of the human race.

BP raised its oil reserve estimate by 1.1% to 1,687.9 billion barrels, which is enough oil to last the world 53.3 years at the current production rates.

The North Dakota, South Dakota, Montana oil shale find in U.S. is 7.5 billion barrels of crude which is less than a 3-month worldwide oil demand of 33 billion barrels per year yet U.S. brags about it like it's the solution to all our energy needs. If we want to keep up with the worlds demand for oil, we will have to find one of these discoveries every 3 months.

When are leaders of the world going to understand if we're going to have a modern society that can last for thousands if not millions of years, we need to preserve our oil for products? We need to quit acting like were the last generation to live on this planet. It doesn't make any difference how much oil we find, we can't drill our way out of this problem.

Man-kinds actual survival depends on our ability to power this planet without fossil fuels in the future.

The new energy invention units are designed to capture kinetic energy from river flow and apply leverage to obtain 1.5-billion-foot pounds of torque for the largest unit designed. We then reduce the foot pounds of torque by 50% for mechanical losses to be extremely conservative. 750-million-foot pounds of torque equates to 1 gigawatt of firm 24/7 electricity.

The distribution capabilities of this new energy invention are just as impressive as the energy projections. There are millions of locations this new energy invention can be implemented around the world.

Think tanks, scientific laboratories and research centers are run by the most intelligent and well-educated people in our world. Any solutions that evolve will essentially be filtered by the leaders of the group. This limits the ideas and innovations to the knowledge of just a few leaders.

Engineers and Scientist also have a built-in deficiency that prohibits them from inventing anything creative. If you bring up a problem to any scientist or engineer, the first thing they do is relate it to some theory they have learned that forces you into a particular approach. Usually, they can explain to you what the theory is and how it has to fit into any solution that you come up with. This is not their fault and even if they attempt to leave their knowledge at the door it's impossible because it's totally ingrained into their personality.

The proof that scientist and engineers have a deficiency in brainstorming is a reality regardless of any arguments people may have. It's been 66 years since we've introduced a new electrical power generation invention. The first nuclear power plant went live in 1951. Since then, we have spent billions and billions of dollars on alternative energy research with not one large breakthrough in electrical power generation capabilities.

Our forefathers from each generation to the next had the courage and fortitude to pass the torch on to the next generation. This is the first time in history where we have allowed old technology (carbon fuels) to be sustained where it is unwarranted due to corporate influence, corruption, lobbying, greed and power.

Our world is stuck in a frozen state and if we don't have the courage to move forward then mankind will face an apocalypse that could possibly destroy the entire human race.

We need to understand the transformations of the next generation will be orders of magnitude greater than our contributions to this world if we just allow science to go forward the same as our parents did for us. They will think we were in the dark ages for using carbon fuels to power the planet.

Humanities Transformation Possibilities of the 21st Century and beyond

The new energy invention that has been created has the capability to generate cheap unlimited clean energy distributed around the planet that will make anything possible anywhere on the planet.

The earth will have no smog, and all energy will be clean eco-friendly alternative energy that will last forever theoretically. Hydrogen and compressed air service stations will fuel vehicles by the electric coming into the service stations without deliveries of oil products. New industries will emerge that recycle all our resources and especially things that we cannot reproduce from oil anymore like plastics, etc. Much emphasis will be placed on food such as fish farms, vertical farms and agriculture to feed the world. The world will eventually clean up all the hazardous sites including nuclear plants, etc. around the world. Oceans will return to their former healthy state where fish will become abundant again. Climate change will no longer be a threat to humanity. We will see a huge effort to restore rain forest and protect wildlife.

Hopefully even a worldwide space program where all the countries come together to build a real foundation for future space travel will see once we solve our sustainability issues. We will then see real scientific research into magnetic drive, plasma energy, 360 water energy, quantum drive, etc. that will eventually lead us to the stars and beyond.

The real question is - has humanity grownup enough to change its priorities to the wellbeing of the human race or does greed and power continue to consume our thoughts and actions?

Worldwide Development Corporation believes this new energy invention will replace all forms of energy on the planet over time including transportation fuels due to its small ecological footprint required and a fraction of the cost of any energy generation that exist today. It's critical if we're going to have one energy source that it's 100% reliable, easily expandable and has backup capabilities. Also, the distribution capabilities of this new energy invention will shock the engineers and scientist around the world. The WDC believes our implementation design will show all the above.

The Leveled Energy Cost (LEC) represents the total cost to build and operate a new power plant over its life divided by equal annual payments and amortized over expected annual electricity generation. It reflects all the costs including initial capital, return on investment, continuous operation, fuel, and maintenance, as well as the time required to build a plant and its expected lifetime. This table compares the US average leveled electricity cost in dollars per kilowatt-hour for both non-renewable and alternative fuels in new power plants, based on US EIA statistics and analysis from Annual Energy Outlook 2020.

New energy invention – Cost per KWh $0.0075 – 0.01 estimate.

The Worldwide Development Corporation value was determined recently to be in the neighborhood of around 15 trillion dollars my claims are true. This estimate was based on the current state of the world resources using a weighted average between the current and estimated replacement cost of the energy generated from the new energy invention. The 15 trillion-dollar value included 5 trillion dollar one time savings for efficiency gains in the electricity grids. We also took into consideration cost drivers that would speed up the implementation of the new energy units. Finally, if we take into consideration without this new energy invention and today's consumption of resources the new energy invention is priceless.

Power plant Type Average Share
Coal 0.1-0.14 0.125 28%
Natural Gas 0.07-0.13 0.1 24%
Nuclear 0.08-0.2 0.14 5%
Wind 0.13-0.25 0.14 1%
Solar 0.14-0.25 0.14 1%
Geothermal 0.05-0.1 0.24 1%
Biomass 0.10-0.1 0.1 0.05%
Hydro 0.10-0.1 0.01 0.05%
Oil 0.08-0.20 0.15 33%

Price differential vs. weighted average $0.141151
WORLD DEMAND (TERA-W)   6.5-6.7tw
20% substitution   1.3
Economic value – per year $1,607,429,580,900.00

I've known John for 5 years and a little more, I can say this with reasonable assurance his invention solves the peculiar problems presented by governments and corporations which rely on pipe dreams to exist. All these intelligent people using this planets resources irresponsibly will be a hard fall from grace when the truth becomes the recipe finality.

I intend to proactively present the future narrative which settles the debate with John Rosebush's blessing, his work as a saving moment in time. What is the working saving us? Simply enough low cost consumer 24/7 energy which solves the consumption of resource, solves the pollution, solves the cost of implementation, solves the maintenance, solves for fresh water desalination, solves for arctic refreezing the tundra is needed when needed, solves for social energy, solves for air quality, solves for trees reforesting, and many other expedience for future needs.

We are at the dawn of new reason and purpose which finds the old ways are dead ways, old thinking cannot respond with proper mitigation needs of the future they are proven not to as we see today solar; wind must be backed up by nuclear or coal !! Which incidentally do harm the environment ecology farming, acid rain and thousands of acres would go to waste under these structures of solar panels. Has anyone understood this yet? Even heat being absorbed radiates horizontal heating the air. We need to stabilize heat cycles not create new ones.

For as intelligent as we are we've lots to learn and fast. I asked in an earlier chapter how insane are we? I mean I've never worried much about the scientists doing good science they do. But I never thought I would have to worry about other leaders and not correcting for error pathways, but today is completely off the chart's ignorance, or plain greed, control, and subjective incompetence to the highest order. What happened to humanity? Is short end thinking the normal now? The future is not lost it's misplaced in old dogmas and old white leaders. Trump and Biden are useless to a proper future.

Young minds must break the false perception these old ways will make a better future it will not, it can't. If you've really read what I've written up to this point you know the reasons and the whys. Now is the time to stand up and be counted as a generation and https://planetaryhealthproject.org [14] is just a stepping stone which points to the correct logical course with viable answers and pathway narratives.

Chapter 9

Richard Kiernicki and TIME-EQUITY

Who is Richard Kiernicki? Why would someone want to change the monetary system? Richard shared with me, "The following quote by Buckminster Fuller gives us a clear indication of why the system needs to be replaced; "You never change things by fighting the existing reality. To change something, build a model that makes the existing model obsolete." Let's face it, capitalism may have worked for the last thousand years, but it does not serve the needs of humanity at this TIME. To a large degree, capitalism, economics and the monetary system collectively are the foundational causes of greed, corruption and manipulation and at present have gone wild. TIME, the foundation of TIME-EQUITY, is fully transparent; it cannot be manipulated, which creates trust and equality, leading to true democracy. The lack of proper financial literacy continues to propel the miscommunication of how the system benefits the few and largely ignores the many. The great lie is exposed, and the system is proven false. Hence, the best way to create long-term positive change is to replace the existing system with a *Star Trek*-like structure, which will release humanity's full potential where there are no longer any excuses to be made that it's not in the budget to do the right thing."

Where does this equity system end up? Who controls this new type of system?

First, Richard Kiernicki is a Canadian and a good friend of mine. We talk all the time about what the future means in terms of current pathways of decision makers. We talk about the realities of finance and in most cases switch to our real interests, helping others understand what our current pathways are defining. Yes defining. We are all subject to the systems now in use, but where do these system head and most of all why do we as a civilization choose these pathways of illogic today? Is it we are not paying attention to consumption and what it represents? Or are we simply out of touch with logical structure of good pathways? Richard opens some doors which I find fascinating as I'm always in consideration of new ideas which find good paths and learning new ideas from those paths. I will try my best to bring the essence of Richard's work in this chapter, from his book "<u>The Capitalists' Worst Nightmare Come True: The Crucifixion of Capital.</u>"

The title might throw you a bit, but it's not a crucifixion of the capital value more it's the Time element which presents the most intriguing aspect of logic to a sensible equality. The intent then is the resurrection of humanity. Time, more important than money? When we contrast the meaning of money and the meaning of time we come to conclusions between different needs or different wants. Let's focus on the needs first. To fully understand where Richard is coming from with the time element, he explains somewhat in dictionary meanings and then his experience with time as a reality personally. The dictionary presents a dry bland time meaning that leaves out the moments of value or changes which accompany time. I'm not going to rationalize the dictionary meaning too deeply as we are learning, right? The dictionary meaning of value really has no meaning as it self-refers to its synonyms. I would prefer a real world tactical and relative meaning for our species understanding as a choice between two abstracts which characterize future use of time as a value of equality. When we begin to define time in the sense it becomes more personal with a care meaning to its value. Ask is your time more important than my time? In the world reality sense how could anyone's time be more important than anyone else's? One can say theirs is more important, but the truth is it is not. We all breath air, we all have heart beats, we all have blood etc.

If any of our life components stops our time here on this world stops. There are no doubt powers which seek to confine our time as seen with covid19, as seen with WWI, WWII and other wars; it's all time for this or that. Who narrates that time use? Who is controlling our time is what we should ask? (When we realize our governments and those who control people in government through manipulative practices like money pay offs and corporate influence, then the time element becomes clearer as an energy consumption motivation and the flow of energy is controlled, then you have centralization and points of time manipulation as tasks for the flow of artificial time.) (The bracketed sentence could be stated more clearly.) This is where we are at today. We do not control our own time therefore out of sight out of mind and out of the learning process which establishes the negative economy or debt economy that can be manipulated. Silent weapons in Time-Energy configurations.

What comes to my mind is TIME-EQUITY has the same attributes except one big difference, the equality factor which checks the implementation because there is no value until the value is created by energy action. People working together to do work and rewarded with their Time is equal to the energy it produces. If this is the case and it becomes accepted by the masses, what happens to the capitalist structure? Does it remain in some form? Is it eliminated to make a new system which answers some of the problems but not all of them with regards to ownership of things? A complete agreement from governing bodies of the idea of TIME-EQUITY would need to be sorted out. Who or What entity would flourish in such a system? If we think further on the social expectations, what jobs are done by who? If the normal business model is no more, what takes control of the time-energy void?

I will refer to Richard's words on the subject of TIME-EQUITY and then I will make sense of the logic and pathway at the end of his writings.

"TIME-EQUITY is the continuous progress of existence and events while being fair and impartial." Richard M. Kiernicki, *The Capitalist's Worst Nightmare Come True: The Crucifixion of Capital* (Toronto: Downloads from Heaven, 2020) 199

Richard explained this as, "everybody's TIME is equal; we all put in the same TIME sharing our gifts and equally benefit from the gifts of others."

"Let's understand one thing, the elite as they are called, the ones who run everything, are finding themselves with too

many distractions and in order to concentrate their efforts of population control they have concluded that the optimum population of the earth is 500,000,000 people. Therefore, according to them, we are in the vicinity of 7.3 billion people too many. You would have to create one hell of a virus and enough vaccines to remove more than 7 billion inhabitants.

Who the hell has the right to make these kinds of decisions and based on what? What education or degree or what accreditations do you need to make decisions like this? Who gave them, and who gave them permission to implement these decisions? If their goal is to de-populate the world it is much more complicated than just throwing out a number; at the very least I would like to know how they determined the desired population. Now they're going to up the ante all in the name of money, you know, the paper stuff, that grows on trees and suggest we keep doing the same thing with a different name. But we're smarter than that now. And we're going to show them.

You must ask yourself who designed, keeps redesigning and refining the rules and laws of capitalism and why? Klaus Schwab, Executive Chairman of the World Economic Forum says, "our old systems are not fit anymore for the twenty-first century," although he enthusiastically supports "The Great Reset" which is the culmination of his 50 years as the head of the WEF.

Schwab's "new" model is referred to as "stakeholder capitalism" which he says is "a model I first proposed a half-century ago, positions private corporations as trustees of society, and is clearly the best response to today's social and environmental challenges." This guy must think the entire world is stupid and that we are going to accept his 50-year plan without question. He earlier stated that what they have been doing "was not fit for the 21st century" and now he wants to implement this antiquated solution which was designed to solve the problems of capitalism from decades ago, including a sprinkle of modern Marxism. Incidentally, the previous model, "shareholder capitalism" has brought us to where we are today, and he concludes that we need "The Great Reset" to convert from "shareholder" to "stakeholder" capitalism; like that shift will somehow now magically change the greed, corruption, exploitation, manipulation and profiteering that created the problems in the first place and then he thinks we are going to accept that this is really going to work this time. Does Klaus really think that he and his capitalist cronies are smart enough to sell us a plan that didn't work 50 years ago and now wants to make it work for the future? Like applying lipstick on a pig conveys the message that making superficial changes is a futile attempt to disguise the pig knowing full well that a pig is a pig. This guy is delusional. That's why he and his cronies need to be stopped NOW!!! Will you be able to live with yourself after? You know and understand there is a better way and do nothing about it?

Have you ever once been asked to join a discussion group with your bank or financial institution in any capacity regarding the lack of equality, fairness, inclusion and social cohesion? Did anyone explain to your satisfaction the rules on interest rates and how they are calculated or the total cost of mortgages? Did they explain the self-serving practices, based upon the majority having to fail financially so the few can really win? What makes you think that any of this will change when their new system of stakeholder capitalism is put in place?

How comfortable are you with your level of financial literacy? Does your level of financial literacy give you a sense of confidence and clarity surrounding any banking or financial questions? Did you ask all of your pertinent questions; and have they all been answered? It seems that a "new capitalism" is once again needed to straighten out the problems of the old system and the economic principles will once again make capitalism the be-all and end-all system all over again. I disagree, the capitalists have had their way for hundreds of years and nothing really changes; they continue to make obscene profits at the expense of their clients and customers.

I say it's TIME to say goodbye to greed and corruption and establish the TIME system for the benefit of everyone's future.

I believe that the powers that have orchestrated the current situation we all find ourselves immersed in. From the first mentioning of Covid-19 and all the uncertainty surrounding the timing, the withholding of true information surrounding the advent of this global pandemic, to the lies and confusion about vaccines and "mandatory" inoculation, to the evaporation of trillions of dollars of wealth as global stock markets crashed along with the removal of our rights as defined by the many Charters of Freedom/Rights, I find myself questioning if I am a conspiracy theorist. Who do you trust?

"The John Hopkins Center for Health Security in partnership with the World Economic Forum and the Bill & Melinda Gates Foundation" hosted Event 201, "a high-level simulation exercise for pandemic preparedness and response, in New York, USA, on Friday 18 October" 2019. "The exercise [brought] together business, government, security and public health leaders to address a hypothetical global pandemic scenario." It featured "a live virtual experience ... to engage *stakeholders* worldwide." Does this sound like someone already knew what was coming? Coincidence?

Bold statements, you need to not only understand the above you need to make a choice and take a stand otherwise you will not be allowed to sit on the fence as your rights are being taken away from you. If you want to remain in control or be part of a different system it is all up to you, the 2%, or more, who are collectively larger than the 1%; please speak up now. I need to know, we need to know that you support a different and humanity focused new world order, one that can be designed with the people for the people as I have suggested in this book. The thoughts that I have shared with you in this book I believe will touch every soul and unite the people of the world in a spirit that is widespread and fundamentally correct in regard to answering the question; "How do we move forward from here?"

"Following the light of the sun, we left the old world behind." (Christopher Columbus)

One last thought to leave you with in this chapter. You may be familiar with Maslow's Hierarchy of Needs, where the bottom of his triangle identifies humanity's basic needs; however, he did not explain the language you use to keep you stuck there.

Maslow's Hierarchy of Needs
Self-
Actualization
Basic Human
Needs
shelter, food, clothing
Words That Descri
love, like, desire, wi
have to
should,
need

(The right side of the above image that is cut off "Describe" & "will")

Words like have to, should and need are the words that support the reality of basic human needs and are extraneous motivators. These words were words used with the best of intentions because they came from your parents, family, teachers, priests and rabbis and others who cared about you and thought that they were just words. But words have meaning and become things. Words that describe self-actualization such as love, like, desire and will are internal motivators and come from your heart. How easy is it for you to act on doing something that you love? Eventually in our individual and collective journeys to the "top," being self-actualization, the language is demonstratively more specific as all of the words associated with self-actualization will speak for themselves as results. Once again, you have a choice, to

remain in the trap of basic human needs or to set yourselves free and be in control of your life, your destiny. It is simple; but not easy.

"Sometimes people hold a core belief that is very strong. When they are presented with evidence that works against that belief, the new evidence cannot be accepted. It would create a feeling that is extremely uncomfortable, called cognitive dissonance. And because it is so important to protect the core belief, they will rationalize, ignore and even deny anything that doesn't fit in with the core belief." (Frantz Fanon)

"You stop explaining yourself when you realize people only understand from their level of perception." (Jim Carrey)

"Properly understood, these changes and their coming have the ability to inspire a degree of hope and optimism unprecedented in the history of your race; for they spell the end of mankind's subconscious condition and therefore portend, as the scriptures of the world foretell, an end to bloodshed, starvation, warfare, exploitation, and needless suffering," Ambrose writes in "*The*

Moment of Quantum Awakening" and continues on, "You will watch as traditions and historical habit patterns once assumed to be survival imperatives are discovered to be detriment to a healthy life and to a healthy society ... experience the interconnectivity of all life."

"You can never cross the ocean unless you have the courage to lose sight of the shore." (Christopher Columbus)

I invite you to look deep within and let go of the past and embrace the challenge of change that we know and understand is needed for humanity to thrive into the future.

Richard M. Kiernicki, *The Capitalist's Worst Nightmare Come True: The Crucifixion of Capital* (Toronto: Downloads from Heaven, 2020)136-144

The words and wisdom of Richard M. Kiernicki

There is the second part of his argument and reason, and I will include it after my opinion on what he has said. First, the reality aspect of a new thinking model will be a marketing affair because we are still bound to a resistive system. When we approach the viable understanding of Richard's intent, we see he is asking for us to believe in a new way of seeing realities. The English language is somewhat limited in its ability to usher the proper feelings in words which can bring that reality next to a decision in the mind of the reader. Richard is seeing the world in a way that most of us do not. His world is a delightful well-meaning embrace of a new pathway narrative which can bring people to that understanding he is visualizing for us. TIME-EQUITY is a natural expression of the rightful comparison to an old worn out system of money exchange. Look at the Romans, they started out with gold coins then silver then copper brass and finally (iron?) IRON. What happened? The system got polluted with greed and all the good stuff was hoarded which left nothing for the state to progress by. Same today only now we have a saturation of paper and a lot less trees. What is it about a capitalist market that fails? It's the belief in one another to hold up their share of the bargain. This being lost creates division. Look at Russia, China, Brazil, they don't want to be party to debt economies, yet they are tied to this capitalism because the leaders must take the lesser of two evils. On the one hand hoarding, on the other hand consumerism. The markets today are rigged for certain timings which allow the hoarders to control more of the consumerists. When this happens there is bound to be a fight because as is today, we are running out of fight material. With their own controls. This would all work if we all had the proper learning. In the next segment Richard provides some of that learning and expresses the will of the people to believe in a better equitable arrangement to solve for hoarding.

The following is from Richard's book and chapter 7, its most telling components.

"What is TIME?

In order to appreciate the concept of TIME we will first have a brief discussion on the information we presently have about TIME.

After we review the information surrounding TIME we will switch the conversation to the knowledge of what we do know about TIME for the purpose of establishing TIME as the equalizer, organizer and synchronizer. Once again, we will commence delving into the definitions of TIME using my standard Oxford Dictionary.

Better definitions and understanding the meaning of the attributes of TIME will help us build a system that is fair to each and every one of us so there is no conflict of interest. Capitalism as compared to TIME-EQUITY, however, is the ultimate conflict of interest as it's all about the material gains at the top and what people will do to get them.

TIME is 1) "the unlimited continued progress of existence and events in the past, present, and future, regarded as a whole." 2) "a point of time is measured in hours and minutes past midnight or noon" 3) "the right or agreed moment to do something" 4) " (a time) an indefinite period." 5) "(also times) a point or period of time: *Victorian Times.*" 6) "the length of time taken to complete an activity." 7) "time as

available or used: *a waste of time.*" 8) "an instance of something happening being done: *this is the first time*" (and it won't be the last TIME.)

These definitions show us how truly ignorant we are of the concept of TIME as we often use the word "TIME" to define "TIME."

A few of my non-dictionary favorite phrases concerning TIME are as follows:

• _Don't kill TIME.

• _A moment in TIME

• _"[T]ime does not belong to us." (Thomas Aquinas)

• _"The trouble is you think you have time." (Buddha/Jack Kornfield)

• _"The greatest gift you can give someone is your time because when you give your time, you are giving a portion of your life that you will never get back." (Rick Warren)

• _"Greater than any spiritual teaching is the gift of time." (Jose Arguelles)

"According to Kabbalah, the gap between cause and effect is called time, and that gap allows us freedom of choice. Because if cause and effect were instantly linked - if every time you hurt someone, they instantly hurt you back – you would quickly learn never to do anything wrong ... In such an

instant feedback universe, who wouldn't be a righteous unselfish caring person? But that would all be too easy, because it would take away free will. Time enters the equation ... now you can't see the cause and effect." (Yehuda Berg)

TIME cannot be fraudulent, dishonestly traded or manipulated, or can it? And here's where the story gets even more interesting.

Sandra Davis shared some of her TIME wisdom with me. "When you understand time, you understand management of processes, production and fulfillment based on law. Time is the law of God for space. It has structure, rhythm and coordination." "The conventional use of time is a misuse." For example, "paying rent on the first of the month." She describes it as a TIME crisis.

This is how bankers misuse the conventional use of TIME to extract additional money from depositors and how and when lenders charge interest to borrowers based on irregular TIME. For example, most people don't know that if you make a deposit to a bank account on a Friday afternoon after 3 PM (or 6 PM in some cases), generally the deposit won't be recorded until Monday, nor will interest be paid. The bank uses that money to make a profit over the weekend. That's why it is better to own shares of the bank and share in the

profits versus having the bank lend your money out to others who need to borrow your money, all the while the bank makes profits for themselves, from your deposits. Similarly, landlords expect payment of rent on the first of each month, which is always on a different day of the week, adding additional stress and confusion for renters.

Further to Sandra's comments I decided to do some more research on the topic of TIME and WoW did I ever receive a mind full.

The floodgates opened and out streamed knowledge about TIME that I have never heard of before. The Law of Attraction in all its glory and a surprise to this author as it added immeasurably to the discourse at hand. At school we were told about being late and that being on TIME was respectful. They taught us how to tell TIME, but they sure didn't teach me much about the history of TIME. Much later as tardiness became a constant in my life, I used the perfectly good excuse that I was practicing for my own funeral. I'm serious, my father passed when he was 56 and I understood clearly that if it was my choice, I did not want to leave this earth before my TIME was up some-TIME in the very distant future. I'm also taking the TIME to live for the many of my family and friends who have departed from this paradise earlier than they may have expected. Would you believe me if I told you that I have even picked my own date of departure?

I'm in love with TIME.

Let me ask you, how much do you know about the subject of TIME?

For example, did you know that our current system of TIME is based upon the 12:60 system represented by a 12- month calendar and a 60-minute clock? Did you know the word calendar comes from the Latin term "calends" which means account book or in other words a schedule for paying bills translating TIME into money? Did you know that the expression "time is money" was first coined in the 1700's?

Our current calendar, the Gregorian calendar, is an irregular standard of measure meaning that the division of the months does not correspond to any cycle in nature. Allow me to explain.

We have months with 30 days, 31 days and one month with 28 (29 days in a Leap Year). This represents an artificial manipulation of TIME in which the existing calendar creates, "considerable confusion and uncertainty in economic dealings and in preparations and analysis of statistics and accounts. The comparability of salaries, interest, insurance, pensions, leases and rents of one period of the year with another is greatly vitiated due to the unequal length of months which have from 24 to 27 weekdays plus Sundays." Once we change the current calendar we improve harmony in government, education, finance, industry, labor, retail, justice, transportation and home-life. [

World Calendar Reform, Communication dated 28 October (1953) from the Permanent Representative of India to the United Nations to the Secretary-General]

In addition to the above, I believe that the globalists have also monopolized the version of TIME that best suits them. This is apparent in the continued use of both the clock and the calendar in its existing form. The current form of the calendar has its roots in ancient Babylon under the guise of The Atlantis Corporation and the "ghost religion of twelve," which "is referred to as a mental disease" that took its form by, "materially incorporating all human needs and services into functions of a privately controlling system called 'money.'"[Jose Arguelles] This will give you a better understanding of the meaning of "wage slavery" and how it occurred and became a part of our lives. "Henceforth, time was to be the pawn of imperial space, enslaved as power units to be known as money." "In this system, money represents power over time" and because money is under strict control and recognized as, "the instrument of power and exchange" the only way to get it is to sell your TIME for money [Stephanie South], almost like selling your soul.

The ghost religion is, "a fraudulent imitation of true spirituality, a false order of reality born of a fundamental abuse of power, and a need to subjugate free will, which is the same thing." This resulted in a loss of equality and was

magnified by the institution of, "taxation and promoted the technology of war to justify its need to collect money and expand its power base." [Stephanie South]

By eliminating money, we reverse this current state in which we find ourselves without explanation, or choice for that matter and therefore will eliminate the need for the war machine as a monetary consideration of economic accomplishment. This does not mean that there will never be any war between people and nations, it only means that wars will not be a fought for profiteering enterprise.

We are currently stuck living in ignorance of TIME's true nature and therefore humanity is living an "error in time," [Jose Arguelles] whereby events and holidays change their numerical dates and/or day of the week from year to year. The calendar only repeats itself every 28 years meaning that it is not an easy task to fix repeated dates.

Since the Gregorian calendar takes 28 years to realign its dates, it is almost impossible to accurately make annual comparisons in accounting for example, as accurate as they should be. Quite confusing wouldn't you agree? What would you say if it was all done to benefit the banks and the financial service companies? It would be interesting to see how the banks increase profit by charging extra days of interest payments in months with 30 or 31 days.

Since the first of the month is always on a different day, "the net effect of the use of this calendar is to perpetuate a fundamental level of mental confusion and ignorance concerning the actual nature of time itself – an ignorance that is hardened into dogma by the unwillingness of habit to consider any other possibility, and to even accept the entire system as second nature." [Jose Arguelles] Dogma meaning, "when you accept something that you never question." [vocabulary.com]

With the Gregorian calendar disharmony is programmed into the system and disconnects us from nature. One might ask why we coordinate the affairs of the world with an imperfect measuring system. So, how did we get to here? We got to here by allowing a small number of others to take advantage over the majority."

Richard M. Kiernicki, *The Capitalist's Worst Nightmare Come True: The Crucifixion of Capital* (Toronto: Downloads from Heaven, 2020)153-160

Okay, Richard has said a lot here. How we are facing day to day without realizing errors are perpetuated by forces outside our control.. !!

Think of this a moment. If we wake up expecting the banks to carry our safety net money and suddenly, they decide its TIME not to carry that safety net because they are self-regulated, which is already here as Richard pointed out earlier, then where does that leave the ordinary people who use this system? The implications are huge to the trusting of this system which could turn on you in seconds. One day you have a nice bank saving for rainy day the next you are empty and no help from the government or local authorities because they are in the same trap.. !

The motivation to change in my opinion from knowing what Richard has said I would be for the TIME-EQUITY system which observes our time rightly as value. If we all stop working for the billionaire today, how would he run his company? If we as a people decided the system is too dangerous to our futures and we stop using services of banks and demanded our time service worth, could these so called billionaires pay us our time? Of course not, the money is all a fake resemblance of our time and our work. What are you worth as a living soul? God tells you you're extremely important, but why? Without the whole truth and whole word of correct thinking we would all perish, not just earth bound but heaven bound as well. Everything we are is tied to that beginning garden events, Genesis 3:15. The Devil has a seed; he was able to use Eve's senses against her and she received from the Devil his life; and what did the Devil's seed do just after being kicked out? Cain killed Abel and so it is today man still being beguiled, still killing and still refusing truth. But there are some who hold to the truth and still establish a correct pathway to follow.; is this where the intelligence of this planet comes from? If we go back into that garden, God says the devil is most subtle, more than any beast of the field. Arrange in an ingenious and elaborate way, making use of clever and indirect methods to achieve something: *he tried a more subtle approach*. What does pleasant to the eyes suggest? Eve was fooled by her senses, and she gave that experience to Adam also called husband. So, when we see how God explains it we understand there are many fitting joints to what man already knew and what God wanted him to know. We can't go back and redo what's been done, and all must see death except those in the last days. Why, what's the difference from the garden till now? Knowledge: how the science will try to reproduce life from nothingness. By the way they are getting close to doing this.

So, getting back to what Richard is talking about as Time and Equity. He is seeing the chance for humanity to take back the redo moment and call out the incorrect and error system and place a correct and solution-based system, which is responsible thinking. If you're held accountable for your errors, you do things differently; this has not happened for the ruling class, and they see no threat of that happening unless you call out the money.

Richard, Signe, John Rosebush and I have created a petition which gives people a vote. Not just the USA and Canada but the entire world will be able to view and decide for themselves if a new directive pathway will be taken, one which is fair and more revealing of its nature as a viable pathway not only in money, but energy and soils as well.

Our reasoning is simple and direct. We solve three major problems when we are all together in the planning process. Richard suggests a single date event on which we all stop using banks and services from these current supply venues and stop showing up for work at the companies owned by the "parasitical elite." It's how the people to take back control. It would coincide perfectly with a transition to the natural calendar on January 1st, 2023, to start the TIME-EQUITY cycle which coincides with the next date the Gregorian calendar aligns most favorably with the 13-month calendar. I've highlighted Richard's work in points which concur with logic and stand on their merits of truth with expressive pathway meanings. To learn more about Richard's work you can email him richard@richardmkiernicki.com or go directly to his website, richardmkiernicki.com/author. I've only grazed the surface of his many other powerful points such as how banks really work, how to dismantle the lie, Time as the equalizer, synchronizer and organizer. The next segment in this chapter is our petition which includes my Soil Remediation pathway to reclaim farming soils' health; oceans waterways and seas stop dying when we get off artificial fertilizers on to organic life-giving means and John Rosebush's water energy invention. I am grateful to Richard to include me in his TIME-EQUITY segments and the petition.

The era of capitalism is nearing its end and usefulness. Why should we be bound to a restrictive 'not in the budget' system?

Why should we as a civilization that's come this far be consumed by greed? Are we brave enough to ask the right questions? Read Richard's work on TIME-EQUITY and you see the beginnings of freedoms which allow for thinking collectively with no restrictions. Image what can be accomplished together.

Agriculture stops dying when we get off artificial fertilizers on to organic life-giving means. I am grateful to Richard to include me in his Time-Equity segments and the petition.

Link   to   introduction   video: https://www.urban2050.org/

UN Petition for a Sustainable Planet

URGENT PETITION

To: António Guterres, Secretary-General of the United Nations

United Nations

405 East 42nd Street

New York, NY, 10017

United States of America – USA

antonio.guterres@un.org

www.sdgcenter@un.org

Our world is on the brink of many systems failing, including our ecological systems (food production and oceans), power generation capabilities and economies...

...the solution is to collaborate in a spirit of peace, harmony and equality...

The UN's states their purpose for, "peace, dignity and equality on a healthy planet." (un.org/en) Let's challenge them to live up to this mandate.

UN Charter Chapter 1, Article 1, 3: "To achieve international co-operation in solving international problems of an economic, social, cultural, or humanitarian character..."

October 31st-November 12th they held the COP26 Climate Summit in Glasgow, Scotland, revisiting old technologies, making promises to reduce carbon footprints, all the while, jet-setting large numbers of people to the conference and motorcades of security, without new solutions for moving forward for a sustainable planet. The usual "blah, blah, blah." John Rosebush of WDC Power, James Boyd Fuller of the

World Alliance for Planetary Health and author of "Out of Balance," and Richard M. Kiernicki, author of "The Capitalist's Worst Nightmare Come True: The Crucifixion of Capital" have been working diligently to bring their revolutionary ideas to the UN to solve the many problems our world faces today. You can help by signing this petition to bring these solutions to the UN that aren't being given the attention they deserve. We need new, truly innovative ideas to create sustainable living and sustainable growth for humanity, not a reworking of the old ideas. Let's make this happen together!

Introduction by Nicolette DeVidar (on the above video), Host of Smart Sustainability on PBS in Washington, DC – "Co-Creating a sustainable, human-centred future. In our own truth."

https://www.urban2050.org/

John Rosebush & WDC Power's UN Challenge:

https://johnrosebush.blogspot.com/2018/03/worldwide-development-corporation.html[1]

"Mankind's ultimate responsibility is to create a sustainable planet for mankind's future because without it everything else is a moot point."

"The Worldwide Development Corporation (WDC) has put together a World Class project to introduce and implement a new energy invention that has the capability to power the planet 1000 times over without consuming resources for a fraction of the cost of any energy generated today theoretically forever. It is the crown jewel

---

1. https://johnrosebush.blogspot.com/2018/03/worldwide-development-corporation.html

of energy generation and the missing piece to the sustainability of humanity."

"The cascade effect of almost unlimited cheap energy distributed all across the planet will be the catalyst to a worldwide 21st century society where there is a place for everybody and a chance for hope and prosperity."

"This challenge will determine if the United Nations is an institution that is willing to lead in the future of humanity or just a body of Nations jockeying for power and influence. I've now added GE and/or Siemens to the challenge if they are brave enough to step forward.

The Worldwide Development Corporation has created an alternative energy invention that we believe will change everything we think we know about energy generation. The new energy invention can generate 100,000 terawatts of 24/7 electricity distributed throughout the entire planet without consuming resources for a small fraction of the cost of any energy generated today. Today our planet generates approximately 6.5 terawatts of 24/7 firm electricity. Even though we would never contemplate or even imagine generating 100,000 terawatts of 24/7 electricity worldwide, I thought it was important to point out the extreme capabilities of the new energy invention."

The purpose for pointing out the extreme capabilities of the new energy invention is that we don't know what planetary challenges humanity may face in the future. The new energy invention can refreeze the poles and glaciers, desalinating enough ocean water to replenish all freshwater lakes and aquifers, power all transportation, generate enough electricity for anything imaginable all without consuming resources and with no pollution. Imagine a world that can power itself for a fraction of the cost of today without any human intervention and that is exactly the capabilities of this new energy technology.

John Rosebush

https://wdcpower.com/[2]

"Our outreach to inform and educate the world's' inhabitants of the real dangers we face and viable answers to the many questions will be addressed ... Our planet's health when we all act together as stewards is obtainable ... The Old energy production venues must be retired to allow the new to be implemented. This new [invention] is extremely essential to the future of humanity ... We need Energy, inexpensive non-polluting and non-resource consuming which equals the future of our species ..." *James Boyd Fuller*

James Boyd Fuller & the Planetary Health Project - World Alliance for Planetary Health:

"In the months ahead we plan to meet and discuss the way forward with leaders to show the importance of good change proper change and technology that will allow it ... The Antarctic is now showing signs of collapse on the western edges with ever more warmer waters reaching the continent undermining the Ice shelves. There is the Arctic, which is gassing hydrates, ocean dead zones of which are growing larger due to artificial fertilizer, rainforests being eliminated ..." *James Boyd Fuller*

https://planetaryhealthproject.org/1st-quarter-2017/

"Out of Balance" a book by James Boyd Fuller: https://youtu.be/jiKbp8cJYkY

https://readerhouse.com/products/out-of-balance/

Mr. Fuller is actively working on healing hypoxic zones in the oceans created by artificial fertilizers and has a solution for the use of natural fertilizers developed from coal by-product's that many countries are currently dealing with trying to eliminate as a waste product. This waste product, when processed properly, becomes natural fertilizer that will help replenish the nutrients in the soil

---

2. https://wdcpower.com/

that provides our source of food worldwide; and stops the needless destruction of the oceans from the toxicity of artificial fertilizers. This project is already underway in select regions of the world and needs to be shared with the whole world so that we replenish the earth for an abundance of food into the future.

What would make all of these solutions easier to implement? What would solve many of the social and economic issues facing the people of our world today, causing needless struggle and suffering?

"Richard is an 'outside the box' thinker and TIME-EQUITY is a bold pathway of change." *James Boyd Fuller*

Richard M. Kiernicki, Author of "The Capitalist's Worst Nightmare Come True: The Crucifixion of Capital":

TIME-EQUITY is a realistic alternative to capitalism that makes obsolete the lies, greed, corruption and inequality largely caused by capitalism. "We have the power to turn the world upside down and free ourselves from the illusions of capitalism forever."

Why should the majority have to fail for the few to win? Why all of the secrets, mystery and lies to cover up how fragile the monetary system really is? Why do some people work three jobs to barely put food on the table for their kids, while others make billions during times of suffering for the majority? During the pandemic 660 more billionaires (google.com) were added to the roster of the rich list in our "Monopoly-game" world, while the governments keep printing money to support the elite agendas?

Why do some people have no shelter, not enough food and no access to clean drinking water, while others live a lavish lifestyle, can eat any food they want, live in gigantic mansions with too many rooms to ever use and do almost anything they want any TIME and with little or no regard for their carbon-producing, jet-setting lifestyles for "business use?" How can we stop the divide between the haves and the have-nots, so that everyone has what they need and has the means to live their true purpose, feed their families and live in a comfortable home, while the world expands and advances without limitations, without a reason not to do the right thing because it's "not in the budget?"

"There are possibilities beyond what we've come to accept as our reality; this comes from an unending drive to imagine beyond the conventional wisdom. TIME-EQUITY is the realistic alternative we've been waiting for to usher in an age of equality, true democracy, a sustainable planet and quite possibly the first era of peacefulness amongst humanity to work together in realizing our unlimited potential."

Imagine a world like in Star Trek, where there are no artificial limitations to progress because of money; because money no longer exists and everyone has what they need, everyone has equal access to resources, everyone contributes to society and benefits from what everyone else contributes.

The opportunities to create new jobs in new environments will create untold potential and even though AI will remove some of the most lack-lustre jobs, humanity itself will evolve with new opportunities reflecting our unlimited potential and the century-old education system will be redesigned to continue to unfold humanity's true creative power beyond our wildest imaginations, to continue to innovate and improve life on earth for each and every one of us.

Richard M. Kiernicki

https://www.richardmkiernicki.com/author[3]

THESE THREE IDEAS TOGETHER – THE UNLIMITED ENERGY INVENTION, NATURAL FERTILIZERS THAT REPLENISH THE SOILS & STOP POLLUTING THE OCEANS & TIME-EQUITY ARE THE TRIUMVERATE SOLUTION TO THE WORLD'S ENERGY,

---

3. https://www.richardmkiernicki.com/author

ENVIRONMENTAL, SOCIAL & ECONOMIC PROBLEMS – THE SOLUTION COMES NOT FROM WHAT WE'VE ALWAYS DONE, BUT FROM WHERE WE'RE GOING...

"Insanity is doing the same thing over and over again and expecting a different result." *Albert Einstein*

"We cannot solve our problems with the same thinking we used when we created them." *Albert Einstein*

Mr. Rosebush has presented his ideas about this new energy invention to the UN before and is in communication with them to present again, along with James Boyd Fuller and Richard M. Kiernicki. The delays continue and the widely publicized disappointment of the COP26 Climate Summit show no new ideas being considered by the UN. However, the TIME is near; let's get this petition out to as many people as we can, let's give the UN a nudge to do something differently than they've done before, to consider innovation over replication and modification of the old ideas. Although this year's summit has already passed, in the days and months following, these issues will not go away if we ignore the real solutions. It is TIME for the leaders to step up, to embrace the logical solutions we bring forward, for the people to show their interest in co-creating the future, for humanity to band together taking on these new ideas and inventions that will solve our world's problems. Together we can solve the problems and challenges of our world with new thinking, new ideas and new-found energy to expand innovate and thrive into the future.

Copyright 2021 John Rosebush, James Boyd Fuller and Richard M. Kiernicki. All rights reserved.

Sign and share the petition. Reach out to us, contribute your input:

https://www.richardmkiernicki.com/contact

Not only are we breaking through existing barriers in these three fields, but we also invite you to solve the other issues (education, healthcare...) and build the team; we invite you to bring your expertise to the table as this is a global project, not for any one individual. There is power in numbers and only together can we grow exponentially and make a real difference in this world. Imagine what we can do when we join together!

Chapter 10

Design of Man is not equal to God &
The mystery of the two Witnesses Revealed

What makes me qualified to make any of the following statements, understandings of revealing's of deep understandings from God or his angel's? My belief started me on a path of discovery, I asked who am I? Why am I? and most of all where is that spirit everyone talks about having that sets them free from the affairs and stress of the world systems?

I found out certain things in our heart must be settled and brought to the table of truth. Everything past and present must be addressed. What was difficult at first was understanding the voice in my heart to my head. With little to no explanation this voice said you are my clock. I was maybe 40 years old at the time. But it was very loud and continuous, I did not understand what that was. Well years later 10-12 passed and the voice said get your house in order and realize your purpose, stop being world lover, idol worshipper and unclean. Those are strong words and convictions. I was reading the whole bible now and finding my errors the many errors of how I viewed God. Like he's there to be God, but I never took me as being in line with him. It was always being in line with man. So that moment I understood this I was directed to the Ezekiel 33:5-12 which made me realize how terrible I was not being with good in all cases. If God can say I have no pleasure in the death of the wicked, how can I possible live with myself believing some are not worthy? That's the day I was born again, I cried now for the remnant outside that garden that needed a righteous prayer for them to allow their eyes to be opened and saved.. !

That's when God or his angel voiced the words you will hear next in this chapter.

This mystery of the two witnesses shows up in Old Testament book of Daniel 12:13 and Revelations 10:8-11 a last biblical book in a series called the New Testament. Gods' own son which endured the system of justice we call laws that ultimately condemned Jesus to death on a tree for doing good acts. Proclaiming if a man looked upon a woman that was desiring in his heart to have her he sinned already. And of Judgment he who is without sin cast the first stone among many more profound teachings.

The prevailing Jews at the time where divided Sadducees and Pharisees of which God was brought into remembrance and against his commandments. The foretold of this very gesture of human against God but in reality, the choice to turn from truth unto fables as in the garden. What steps out and takes shape is the uncanny accuracy of the Old Testament guiding the new. How? Ezekiel 33:5-11 instructs those who believe God to warn them to stop doing evil against one another or their blood will be upon them. But God goes further to say he has no pleasure in the death of the wicked, only they change to good and be saved from eternal separation. When Jesus came proclaiming the letter of the law more precisely if you look upon a woman and in your _heart_ desire her you commit the sin. This sort did not set well with the ruling class of the day and I'm sure it still does not today.. !! The flesh is weak and subject to pains and desires. The _Times_ dispensation begins by spirit this time and explicitly says grieve not the holy spirit whereby you are saved. Not to believe isn't the sin this time it's the intent of why you don't believe is the separation as in have a form of godliness and deny the power of God.

We are instructed to believe and have faith in what Jesus has established for our good. There is that choice to believe or not believe. What is about to be established is why this has come to pass and is on the collision course with an authority taking power over the earth. In the end Evil will be done away with for 1000 years and the earth heals and the believer reigns over the earth because evil is locked away and sealed shut until a final judgment is proceeded against that evil as testified by you and me and all the other witnesses throughout history, and yes future. Do we know of other dimensions? Other realms? The word of God says they exist and science is now hot in the pursuit of the mathematical values needed to support such thinking and realities.

To establish truth which defines why we are in this testing state of biological carbon and who objected to such creation of man and women in Gods image is our next understanding.

First, we ask the needed questions which show intent and action from that intent.

In the *Garden of Eden* how many where in that garden? This is important to know because it sets the stage for the next 7 millennia of time as a separation from God by choice.

If you said two that is wrong. If you said three that is wrong. If you said four that is correct. There were four personages in that garden. My next question is why a garden? Because outside that garden was violence and evil the creation of Lucifer the fallen angel which accuses God and heavenly host of not telling the whole truth in heaven! Lucifer was cast to the earth looked up wondering where he was. He was changed to lower state of being a man. In heaven he was the accuser Revelations 12, so God dealt with him by casting him out removing his power changing him and placing him in the Garden!! For the benefit of life. The devil as the tree of knowledge of *good and evil* He knew both avenues and accused Gods host of creating man in Gods own image that its wrong for God to do this and not covering the whole truth. Lucifer now the Devil the creator general and refused to relent he thought his creationism was total and above God's word. This leads to what God did next. He made man from the dust of the earth and presented Adam to do the earthly naming which ticked off the devil even more but could not touch

Adam for the devil was made a man and subject to human failings and weaknesses but had full remembrance of heaven his intelligence remained. After so long God saw Adam needed a helpmate and took a rib of Adam and created the women. This tempted the devil because he knew evil.

I will ask one more question, why did God put the devil in that garden? If you say to be tested to repent of his evil, you would be correct. We all know there was the _Tree of life_ in the garden also known as Christ the personage of God which can forgive sin even the separation sin. But what did the devil do? He beguiled Eve and Eve gave also to Adam. What did she give Adam? She gave Adam the way to make a baby their own flesh reproduced. This is what separated man and the devil from God until the _End of Days._ That end of days is here as promised from that word of God to Daniel the Old Testament prophet and revealer.

Lucifer became Devil at this time because the act of removing Gods creation from belief required a sacrifice and the tree of life was the witness to the act and became that sacrifice, we call Jesus Christ in the TIMES dispensation.

While I was reading and writing another book about nature and the out of balance state that is become obvious, I was drawn to the book of Daniel chapter 12. An inner voice said write the words which I will tell you for they concern the end of days and my two witnesses in spirit and in truth.

I was directed to Daniel 12:7. And I heard the man clothed in linen, which _was_ upon the waters of the river, when he held up his right hand and his left hand unto heaven, and swear by him that lives forever that _it shall be_ for a time, times, and an half; and when he shall have accomplished to scatter the power of the holy people, all these _things_ shall be finished.

To those who read this revelation made known mystery of the two witnesses will wake a fire in the mind that grows the tabernacle of *peace patience* and testimony of the living promise of the living Word of God. Through the spirit of Jesus Christ to make known his will for the end of days. Repent be baptized of the Holy Spirit in word and promise by all power and authority, amen.

All one must do is believe in Jesus Christ the Son of God and do the 7 spirits of God. Love, Peace, Truth, Charity, Compassion, Strength, Power.

Before beginning the next part, I need to establish the context of the preceding paragraph. Peace patience is those observations of Ezekiel 33:5-11 which places blood on the hands of those who would do evil to another human being. How to deal with their condition by Matthew 5:44, Luke 6:27-28 and several commands of this effect or intent of the heart. Pray for your enemy, misguided leaders, deceitful, and those who are wicked. On prayer for them rests all the law and the prophets which made known these understandings. Put it this way many of these people are in bondage to blindness and our love for them can set them free of those bondages. Imagine if you were headed for certain separation from your family and no one helped get you, your family back, how would you feel? The law and the prophets are our guide back to truth which settles our pathway to know good and correct the errors.

Of course, the authority of this world does not like you taking his souls he worked so hard to steal, yet our power in prayer is over such evil by promise when you follow those words God has presented to us as a way back to his realm. He promises no more tears, pain, sorrow, or separation again after our testing is complete. When we practice good, good comes in the end fully manifested. Death is not the end.

What is in a number to Gods plan to save humanity from EVIL?

Who is supposed to recognize the evidence that wakes up the sleeping?

Why does the Old Testament have a LAW?

Why does the New Testament provide a 7-fold MESSAGE?

[Red Heifer] born 28 Aug 2018 this allows the false Jews (Rev 3:9) to build a third temple and satisfy their doctrine in old law becomes a sin Numbers 19.

....

Daniel and John,

Daniel to sleep until the last days Daniel 12:13 and John to prophecy again in the last days. Rev 10:8-11 God Moves forward.

First, Daniel 12:7 And I heard the man clothed in linen, which was upon the waters of the river, when he held up his right hand and his left hand unto heaven, and swear by him that lives forever that *It shall be for a TIME, TIMES, and an HALF.* And When He [Satan] shall have accomplished to scatter the power of the Holy People; all these things shall be finished. Notice: ALL these things will be explained.

The words that be established hereafter are the culmination of the indictment of Satan as his refusal to repent and turn from evil unto good as testified by John in Revelations 12. From Rev 4 to 19 tribulations to turn evil to good and the transition of error to correct thinking being refused.

The setting up of the abomination that makes desolate is the primary focus of this writing to establish the Time of Many and the

Word of Truth to establish testimony and to establish witness reality in purity and sacrifice for the living and the dead in Christ.

Note

Time=Old Testament 1954 years 42 generations 33AD

Times=New Testament 1954 years 42 generations + 70 weeks of Daniel=2024

Half=End of Days 2025-2032 during the week 3.5 oblation ceases 2027.5

Daniel 12:9-13

9 And he said, go thy way Daniel: for the words are closed up and SEALED till the time of the end. *Rev 10:7 declaration through a prophet 7 CHURCHES, 7 SEALS revealed.* Dispensation of TIMES.

10 Many shall be purified and made white and TRIED; but the wicked shall do wickedly: and none of the wicked shall understand; but the wise shall understand.

11 And from the time that the daily sacrifice shall be taken away, and the abomination that makes desolate SET UP, there shall be a thousand two hundred and ninety days. *This is to say 3.583 years.*

12 Blessed is he that waits and comes to the one thousand three hundred and five and thirty days. This is to say 1300 and 5 and 30 days. Remember this is (waiting) this is another group of people and or in other notation 3.70.

What Group of people? Rev 22:17 The Bride of Christ! The Elect the No Guile The Meek The peacekeepers. How will they act? They EDIFY not blame or provoke or Lie.

13 But go thy way *till the end be*: for you shall rest and stand in thy lot at the *end of the days*. Daniel will return to his lot and declare Gods truth and Repentance for the Jew, both of them. The true Jew and the false Jew Rev 3:9-10.

What is Daniels Lot? [The true Jews] He will speak again to them.

God declares an end to days. God also declared two numbers of which will define the end days reality to the bride and those that love God and follow the Lamb.

There is no mistaking Time, Times and a Half for Old Testament, New Testament and End of the Days Testament Witness.

Those who oppose this testimony are the lukewarm and evil ones; they have a form of Godliness but deny the power of Gods words. For it comes from an unlikely source and God knows the struggle of the pure at heart. How truth is suppressed for the gain of a lie. Thus, this generation is at the second *Times and is already underway with the Half and is ready to be accomplished. Thus, will Daniel rise and see the end of the days. The Old Testament must be acknowledged as a witness therefore the first Time is the allotment. Daniel is the Revelation of the Old Testament he delivers the message to the true Jews.*

Rev 10:8-11 The little book is John's testimony of the end of the days, which will indict Satan before the council of God as the testimony and witness of the people and their suffrage and earths degradation.

The little book also presents the testimony of the spirit of the age as bitterness of bondage to deception and lukewarm spiritual condition. The new prophecy will be in numbering TIME, TIMES and HALF. Half is the End of the Days spoken by Daniel

TIME

The numbering; Matthew 1:17

Time=1954 years and 3x14=42 generations as declared in Matthew 1:17 as also outlined for our generation 2017 the start of the setting up of abomination that makes desolate the computer networks, the building of the third temple in Jerusalem and red heifer being born with no blemish no yoke numbers 19. *A circle of Time*, which brings Gods remembrance of Evil. Jesus Christ is the third Temple!! Man is now adhering to fables and unclean living both in nature and values of moral standing where the love of money is regarded above God.

A single generation therefore is _46.523_, which TIME is set for an age of God with a people. Jew or Gentile. In our case the Gentile makes a name for Jesus Christ.

In this case the *LAW* of Moses out of Abraham and Melchizedek. The Jews are the chosen and for 1954 years God is with them. Then they DIVIDED [*Sadducees, Pharisees*] cause God to be in remembrance of Evil for that age. God warned the TIME through Prophets to not practice idolatry, yet they do and did. Therefore, God set up TIMES another beginning age with the Gentiles through Jesus Christ Gods own projection of Truth, Peace and Moral behavior Mercy, Grace a law to itself.

TIMES

TIMES=1954 years as the Old Testament the New Testament is similar but with key differences. 7-fold declaration of Jesus Christ, 7 church Ages 7 seals, 7 Plagues 7 Vials and a Prophet to each age as defined in Revelations Chapter 2-3. Revelations 10:7 to make mysteries known to the _7th church age_ as a warning of the entering into HALF to REPENT be baptized as in the day of Pentecost. [This is where the disciples waited for Jesus as instructed and visited them for their belief in their patience; CLOVEN Tongues surrounded them baptized them with fire] The way of man distorts Gods message of LIFE, PEACE.

It's extremely important to know this *End Days* message correctly and God warns through Matthew 24 in Jesus' own words the progression of MORAL DECAY and yet there are the watchers [Joseph Smith LDS & John Wesley Philadelphia] 6th church age [Philadelphia] who keep their Moral truth unto death and the constitution of the dead to be saved. Like the four horse riders the delegation of times is specific.

Rev 6:1-8 White horse rider with bow no arrows [Devils Doctrine no power] Red horse rider has power to take peace away and enforce it with a large sword [doctrine has power takes peace from the earth] Black horse [Authority exercised] Pale horse [Death & Hell] These repeat throughout generations except the last.

The 7 seals book is opened by the Prophet <u>Rev 10:7</u> of Laodicea last age of which was William Marrion Branham 777 to make known what those seals are and how they relate to the purifying of the bride souls to Christ. These carry power but are overcome and taken from the earth to meet in the air. Prayer, a standard against evil, is eliminated by Satan through separation.

Now the lukewarm are informed follow Satan or die 2027.6.

Branham had the attributes of a prophet which is <u>discerning the hearts</u> telling the people what they are thinking, <u>healing</u> he was a healer through Christ, casting out devils, he could cast out evil spirits, and as hard as it is to know the truth he delivered that truth, many did not like his messages none the less he did deliver the mysteries and opened the <u>7 seals</u> said there will be another final messenger to the Half age.

As put forward in Daniel 12:7 When HE shall have accomplished to scatter the Holy people, all these things shall be finished.

[HE] being Satan. All these THINGS, what things? The Declarations, the 7 church ages the 7 seals, the building of the third temple in Jerusalem, Satan destroying the earth through man by technology 5G mm wave network and deceiving the people with false miracles and money more important than helping the neighbor in time of need. 5G mm wave Network is the abomination that makes desolate the physical breaking.

Revelations 10:8-11

8 And the Voice which I heard from heaven spake unto me again and said Go and take the little book which is open in the hand of the angel which stands upon the sea and the earth.

9 And I went unto the angel, and said unto him, Give me the little book. And he said unto me take it and eat it up and it shall make the belly bitter, but it shall be in thy mouth sweet as honey.

10 And I took the little book out of the angel's hand and ate it up and it was in my mouth sweet as honey and as soon as I had eaten it, my belly was bitter.

11 And he said unto me, thou must prophesy again before many people, and nations, and tongues, and kings.

This is Gods proclamation setting up judgment to the rulers of the nations, rulers of the churches, rulers of the people and authority of testimony.

John will prophesy again!!

This means the spirit of John is with us in flesh on this planet, but where and when? John is a Jew a converted disciple of Jesus Christ he will therefore be in Jerusalem as one of the two witnesses. Daniel is the other witness, his standing in his lot! Daniel 12:13. These two will witness and prophecy for 3.5 years 2025-2027.6 and they have power to shut the heavens that it rains not and power to send pestilence disease to the wicked and unbeliever among other things.

I have prayed to God I could see these two and ask blessings in their sight for the bride and the vesture and their protection toward those who love Jesus Christ.

John's particular testimony is Gentile oriented also for John a Jew. Johns' prophecy for 3.5 years Rev 10:8-11, the bride is gone up after 6 months of the start of tribulation sometime near the end of 2024 the setting up abomination that makes desolate will complete with NO CHOICE. The 5G network and Miracles of Fire, Images in the AIR. This is a Coup de Grace to peace on earth and peace is taken from the earth, no more prayer the standard against evil is done away gone up to God as their reward their days shall be shortened.

After the sealing of the Tribes of Israel 144,000 there is another great sight in heaven Revelation 7:9-17 These are the left over from great tribulation That made their garments white with the blood of the lamb and got victory over the beast and false prophet and Satan. This is not the Bride or the Elect, these are the lukewarm that realized they were deceived, repented and called on Jesus Christ for mercy for mercy is still active after the bride is taken off the earth. Grace is given to the Elect and the Bride the Holy are gone by chapter 3 of Revelations. The Lukewarm go through the entire Great Tribulation 7 years where Jesus Christ puts an end to Devils and locks up Satan the dragon, which deceives the world. We are the Witness those that practice Love, Peace, Truth, Charity, Compassion, receive Strength, Power in righteousness.

Malachi 1

4. Whereas Edom say, we are impoverished, but we will return and build the desolate places; thus says the Lord of hosts, They shall build, but I will throw down; and they shall call them, The border of WICKEDNESS, and The people against whom the Lord hath indignation for EVER. Why? Because they do not believe.

This is none other than the promises rejected of these false Jews and them joining to reprobates of idols of stone and wood that cannot speak nor walk or produce truth.

These are the false Jews that cannot find the truth when God plants it in his prophets.

This is the ending of God dealing with Evil to turn them from Evil. And its forever.

This all sets up what God will do next. Send his son as a sacrifice and a beautiful offering of his own life to save the corrupted souls through choice by Grace and Mercy as promised. Ezekiel 33:5-12 we are to warn, and God takes no pleasure in the death of the wicked, only they turn from evil to good.

Church age Messengers, 1 Apostle to Gentiles

Apostle Paul Ephesus 53-170 AD The Apostle to the Gentile dispensation.

Ireneus Smyrna 170-312 Ireneus Knew Polycarp from Smyrna he was martyred for Christ.

St Martin Pergamos 312-606

Columba Thyatira 606-1520

Luther Sardis 1520-1750

Wesley Philadelphia 1750-1906

Branham 1906-1964 LaoDicea Said there will be one other to declare End of days witness testimony what it means. Where the eagles will gather. The eagles gather in the air 2027.6

2025 John & Daniel/End of Days 2027.583

Martyred for Christ and his people.

Daniel 12:10 Many shall be purified, and made white, and tried; but the wicked shall do wickedly: and none of the wicked shall understand; but the wise shall understand.

Revelations 22:11 He that is unjust, let him be unjust still: and he which is filthy, let him be filthy still: and he that is righteous let him be righteous still: and he that is holy, let him be holy still.

Revelations 22:12 And behold, I come quickly; and my reward is with me, to give everyman according as his work shall be.

Revelations 22:17 And the Spirit and the Bride say come. And let him that heareth say, come. And let him that thirst come. And whosoever will, let him take the water of life freely.

Revelations 22:21 The grace of our Lord Jesus Christ be with you all. Amen.

As touching on the Red Heifer

What is the Red Heifer? What's the significance of this happening now? Why should this matter to any of us non-Jews?

First of all, the Red Heifer is a red colored cow which God set in Number 19 of the Old Testament bible to cover the sins of the people as a sacrifice. What is peculiar now is that there has not been one born in over 2,500 years which meets the requirements laid out in Numbers 19. Coincidence there is one now born in Jerusalem August 2018?

The significance of this is threefold in determinations. One is the False Jews will be allowed to build a third temple. The false Jews will think this animal will remove the sin of the nation and Israel will be back to God like in the beginning. Three the setting up of abomination that makes desolate the authority is given which is supposed to place power in their hands. When this animal gets sacrificed these false Jews are out of the law as the practice ended after Jesus Christ made perpetuations for sin. Daniel 9:24 Seventy weeks are determined upon thy people and upon thy holy city, to finish the transgression, and to make an end of sins, and to make reconciliation for iniquity, and to bring in everlasting righteousness, and to seal up the vision and prophecy, and to anoint the most Holy."

Those 70 weeks are a prophecy which unfolds the sealed vision in the end of days which is opened in revelations 10:7 and in 10:8-11 John is instructed to prophecy again before many nations' kings and people. What is it? It's the final choice to accept truth or reject truth. God makes his intentions known and here we are looking at the reality in that Red Heifer being born August 2018, Coincidence?

Will it be a coincidence when these false Jews sacrifice the animal in August 2022? Every word of God will be established and then will the end be. So where are we in Gods time? Not man's time Gods time. We are two years to the absolute separation of truth with fables as God puts it a remembrance of Evil a circle of time. Will there be the saved reaching out to warn the people where they're at? God always warns his people. I am one person among millions of people saying the same things I am, except God is giving the times the years and his absolute love in so doing this for those who need their chance to make a better choice than the Devils choice of lust, greed, pain, suffering, diseases and lies. Our day is arriving, do we know the day or the hour, NO. Do we know God will do what he says he will do, yes in my case I stand with truth and with Jesus Christ because my learning has brought me to this place as a servant to make known what he said declare?

Many will question the validity of the words written and find fault or a way out of the intent of this writing. Some will say you can't add to the bible in revelations and change the meanings. This is true and I have neither changed the intent or meanings, but the spirit of Jesus Christ has made clear a timeline no one else has understood until now. As for the day or hour where have I changed any day or hour? The spirit presents the Years these will begin to happen not the day or the hour of its completion.

The message is clear repent of evil, turn aside from mischief and uncleanness. Open the heart to truth and realities will unfold which promote good to the fellow man, woman and all neighbors as our testament and witness of the truth by actions.

When we look at the establishment of the Law out of Egypt what is the purpose? What is the intent of that Law? According to archeological evidence approximately two million Hebrews were released by Pharaoh. These made their way to the Sinai desert in which the law was established through Moses. What is peculiar about this story is two of the two million of the original group made it to the land flowing with milk and honey. This was a promise from God this would be the inheritance and posterity of the Hebrew nation. When in the promised land who were those two that made it? And why?

Joshua, the second in command to Moses as touching the spiritual and Caleb as believing God and his good report of the land he had witnessed. Why is certain, as being an intent in the hearts of these two. But why did they make it? They made it because even at the odds against them and troubles encountered, they did not waiver in their trust in that promise. So, it is today that same trust must be steadfast through all the pains and seemingly endless hate and hurt we all endure daily on this planet. We have a place prepared for us that's a promise from Jesus Christ, will we honor our purpose our choice? When we do how do we know? When we honor that purpose God is bound to his word and will give you what you need to make that pilgrimage, that transition to everlasting.

There are spiritual aspects to life we have a unique perspective on what is unfolding around the world in plain view of all the people because of the internet. TV, Magazine, media moguls, bloggers etc. There is an attempt right now for good people to find solace and peaceful futures. All the people today are not bad people they are being told to divide or separate from those who do not think like them. This is wrong and the spirit of truth tells you it's wrong, that is that voice that speaks the word of God in your mind to turn the corner and change with that turn which puts you back on the right path. Ask did God cause the wars which man fights? No. What caused the wars? Separation, division causes war. How can another human being cause judgment on a righteous soul? This is the essence of our plight. When we talk of justice, what justice? In the word of God does he say kill others? Does he exclaim those humans are worthless? So why the Law? Because man is unable to correct in spiritual matters which get him back to where we are supposed to be, observing Love, Peace, Truth, Charity, Compassion which bring you to Strength and power of LIFE.

How do I know this? How can I be the only one which narrates a proper future? The answer to these questions in practicing the spirit daily and believing the original narrative from the creator. Why does the creator love his creation so much to give a battered do-good spirit as a sacrifice for many? Remember this Jesus was killed for doing good deeds. The world has not changed that attitude, but one thing is now prevalent that we are all in the knowledge of such happening and whether we believe or not this creator is still offering a way back to his world that we were all kicked out of and changed to be given choice by a more perfect knowledge.

The deep things of God what are they? Where do they take us? How do we know? Why does God stay with a single voice for revealing his redemptive path at certain times of need?

First, let us search history and find truths there. This will provide a baseline for how civilization acts when given more knowledge from prophets, teachers, holy people that *regard life as precious.* Throughout the ages the narrative of life in the eyes of the creator of heaven and earth has followed one predictable pattern, to turn the hearts of the fallen from grace back to him by their choice. What was happening in heaven which prompted a serious change? From what we can glean in records from early civilizations is there was a war. Now war in heaven a created place of God must have been a pretty big deal because God created another place to contain the souls of the war makers and try to salvage them by giving them another choice another chance. The earth was created and was without form and void meaning nothing alive could live there. As we know being science based along the way our human capacity makes earth some 4 billion years or more old. Is God the creator of this universe 13.7 billion or more years old? Our science instruments are limited.

Remember God created heaven and earth for what? Was the tree made for the earth or the earth for the tree? So, we cannot say how old God is or how he came to being he just said I am that I am. When we take near history Phoenicians and older such civilizations, we are always finding new evidence of even older civilizations which had remarkable building skill of which would be hard to match today. So, what went on during these periods? When man first arrived from the biblical perspective first there was war in heaven that caused a separation, this separation began with an accusing of the heavenly host of something which caused Michael an arch angel to stand up and his angels and defeat the dragon or a mighty creator angel given power to create. We ask now was that 4 billion years ago. What if it was? Does it matter that the age of the earth from creation point till today? What truly matters is that bit of information which tells the story of the war it did happen and for the accusing, What? This is one of those deep understandings from God. The Satan was against the creation of man in their own image as Satan put it God lied and did not tell all the truth leaving Satan a wanting creator. The accusation was selfish and had no meaning. This sets up what God would do next. After Michael and his angels defeated Satan, God said let's create a garden and have man in the midst of that garden to keep and dress it, let us form him of the dust of the earth and give him a *Soul*. Let's stop right here.

Let's answer the question of a "soul." What is it and why were we given one? God does not talk much about why man has a soul, but he does say it's good all of his creations are good. What is good? Good does not harm and does only edify reality to keep good in balance. God wants the very one that caused the separation to become good again and so in that Garden of Eden the devil was placed changed into the very thing God had created from the dust of the earth. We now ask who was being tested in that garden? The Devil.

We know by reading the account of the garden that God also made Eve from the rib of Adam for a help mate after which he instructed both of them to not talk or be part with the tree of knowledge of "*Good and Evil*," the truth of which God had changed and made him mortal forever. We read revelations 12 and find this war described by one of John's visions, the part that tells the inhabiters of earth the Devil is come down unto you full of wrath, what does that mean? He's pissed and wants revenge, but for what? So we have a time line set up which there are inhabitants on the planet. This is the *"casting of souls."* Casting of souls is the requirements of good upon all the fallen angels which sided with Satan and his fall they are outside the Garden and have knowledge of heaven but no power to exercise angelic authority as service to God. Now the reality of the garden is this. When the Devil understands, he is cast to earth out of his heavenly realm he is cast on the earth and changed but his memory is intact and does not allow him to change that mind set unless repentance enters into the understanding. This is why God calls him the tree of knowledge of good and evil.

Now we must put into perspective the reason he is in that garden to begin with. Being made mortal he is now just like man in all respects even hunger. We must also show the entirety of the scene, so we ask how many are in the garden? Turns out to be Four as explained earlier. But what are their purposes? Adam, we know is a dresser of the garden to keep it Eve is the partner to Adam. Then there is the Tree of Life which is Christ the personage of Good and bearer of life everlasting and can forgive sins or other holy duties to the protection of life itself as God is love so too is Christ the bearer of that Love and sacrifice for life and its continuance. Then we have the Tree of knowledge of good and evil, what's he doing there in that garden? He is being tested whether he will ask forgiveness for sin against life and repent to Christ the personage of God. When we watch the story go further, we see the devil uses the women and beguiles her and wakes her senses. She is overcome by the devil, and she gives also to Adam for she had two children and the one killed the other, is that right? So, it is this day war with the women's seed you and I are that seed. But there is an end, and those souls are carried throughout the ages to be tested and find good by choice just as the Devil had choice. When we read the finality of the matter, we find the women or the church is given a resting place where through the ages the church is responsible for making known truth and life and peace and there is a Time, and Times, and Half a time from the face of Evil to be perfected in Christ, but one must believe with the faith of a mustard seed.

So how do we know all these things had taken place? When we practice the seven spirits of life, we are promised we will see God and do what Jesus did in his day. The blind sees, the lame walk even cancer cannot stand in the way of righteous prayers for God is life and commands the evil to depart from a body. So, when we are able to recognize the inherent truth from god to us we too command life with word and promise by the spirit which performs the need and sets reality in place for all to see it and believe also. But why do we not have this power always? We are soul with flesh which is susceptible to beguiling spirits The parts that Satan can command and promote is doubt and other knowledge which skew the truth. There can be thousands of books written about doubt and science or more precisely pseudo-science which perverts the intent of the original words and find reasons not to believe not to have faith, and not to live a better life which upholds the principles of Christ and life. Therefore we have disagreements, wars famine pestilence and covid19 mRNA gene therapy which cannot in the least help the condition of disease, man made it man will die by it. Man thinks he's at the pinnacle of knowledge and proper thinking when in reality he has reached blindness, naked and don't know it. What a terrible condition to be in.

Here is an example of thousands I encounter, but this is a well-known reporter intellect and generally good person who faced what many can only speculate at as the reality points of the deep things in life which are meaningful and faithful. The following is in the words of February 18, 2022, | John Horvat II

Dr. Naomi Wolf is not given to wild imaginings or conspiracy theories. She graduated from Yale University and received her doctorate from Oxford. As a bestselling author, columnist, and professor, she is outspoken in her defense of all leftist causes (including procured abortion) and is not afraid to say what she thinks.

However, she recently encountered the "principalities and powers of darkness." Her recent article, "Is it Time for Intellectuals to Speak about God?" describes this discovery process in brilliant detail.

Experiencing the Wrath of the Liberal Establishment

Her case triggered this wrath because it threatened a vital narrative crucial to a progressive[1] agenda. (*Hmmm wonder what that is? Could it be silent weapons for quiet wars*) It need not be a single issue. It can be the pandemic, the vaccine, the economic meltdown or breakdown of order. Indeed, conservatives have long experienced the disproportional wrath of the opposition on a whole spectrum of issues. What is different today is the increased intensity of this wrath and the willingness of liberals to attack their own.

Dr. Wolf's crime was questioning the effects of the vaccine on young girls. She has long produced similar exposés denouncing big business interests, all to the applause of the liberal establishment. She expected the same reaction to her well-researched articles.

However, she was shocked to find herself on the wrong side of the narrative. She soon discovered that life-long well-educated colleagues, journalists, editors and professionals—critical thinkers all—rejected her broadside and clung to the talking points of CNN and other major media. They asked her not to show them the evidence lest they lose status, jobs or opportunities.

---

1.   https://www.tfp.org/progressive-clergy-intend-normalize-sin-sodomy/

"The progressive, right-on part of the ideological world—my people, my tribe, my whole life—became more and more uncritical, less and less able to reason," she exclaimed as she found herself "pulped, de-plat formed, canceled, re-canceled, de-plat formed again, and called insane." She was accused of being a "pandemic conspiracy theorist" engaged in "crack-pottery."

The issue was not the vaccine but the challenge to the narrative. Indeed, *The New York Times* broke the same story in January 2022[2], vindicating research similar to hers but not reestablishing her reputation.

Evil Too Massive

She perceived that the broadside was not against her but represented "an infection of the soul," the abandonment of "the most cherished postwar ideals," and the dropping of "post-Enlightenment norms of critical thinking." She was impressed by its orchestration and agility.

This overwhelming attack represented something profound—and sinister. Dr. Wolf concluded that "this edifice of evil is too massive, too quickly erected, too complex and really, too elegant, to assign to just human awfulness and human inventiveness."

She was perplexed by this intelligent force that was much more powerful than anything she had ever experienced. She recalled the comment of a Christian medical freedom activist who had suffered similar treatment and calumny. He said he found courage in Ephesians 6:12:

"For our wrestling is not against flesh and blood; but against principalities and powers, against the rulers of the world of darkness, against the spirits of wickedness in high places."

When the hammer of evil fell upon her, something clicked that made those words come alive.

The Necessity of a Countervailing Force—God

2.   https://www.tfp.org/a-not-normal-2022-means-its-time-to-improvise-and-dare/

Indeed, she had encountered the devil[3] who embodies evil in its most extreme and radicalized form. She felt "the majestic nature of the awfulness of the evil around us, the presence of 'principalities and powers'—almost awe-inspiring levels of darkness and of inhuman, anti-human forces." This narrative generated "anti-human outcomes" beyond the reach of humans alone.

From this sinister impact, there awakened in her soul a belief in "the presence, the possibility, the necessity of a countervailing force—that of a God." An evil so big demands a God that can address this danger.

With her tiny faith in an unknown God, she started to pray. She might well pray with the disciples on the boat in the raging sea who awakened Our Lord saying: "Save us, Lord, lest we perish!" (Matt. 8:25)

A Leap of Faith

This literal leap of faith[4] presented Dr. Wolf with many problems. She expressed consternation because her Jewish cultural background does not frame spiritual battles in the dramatic fashion presented during her recent ordeal. Classical liberal authors like her, writing in a postmodern world, are not supposed to talk about God, much less the "principalities and powers of darkness."

She reflected that the rise of Existentialism after World War II glorified a worldview that celebrates the absence of God and humanity's essential aloneness. Thus, few dare talk about anything spiritual.

However, she resolved to talk about spiritual combat and "the fate of souls" because "the forces of darkness are so big that we need help"; humans alone cannot solve this predicament.

---

3.   https://www.tfp.org/the-devils-false-promise-of-happiness/

4.   https://www.tfp.org/society-based-faith-charity-look-like/

She does not know how to address this God or even describe Him. All she knows in her primitive faith is to call upon this unknown God who represents the only hope for an evil world.

Lessons From the Case

The case of Naomi Wolf is a sign of our sinister times. We do not know whether she will develop her faith and embrace Church teachings. Such conversions are best left in the hands of God, whose ways are mysterious and inscrutable. However, her desperate plea for help against the principalities and powers contains the seed of faith that may blossom one day.

The most encouraging conclusion to this story is that sinister times call forth courageous leader souls who recognize the nature of the evil we face and listen to the voice of grace[5]. As Saint Paul says: "Where sin abounds, grace abounds much more" (5:20). How many thousand Naomi Wolfs are out there that are finding God amid their desperation? With their little faith, they are calling upon a God that they never knew and now grasp. God cannot fail to listen to such pleas.

The case also contains lessons for mediocre believers who do not want to see the spiritual combat and make no effort to battle the principalities and powers. Her cry might also awaken these lukewarm souls. They might pray: O God grant me the small faith of Naomi Wolf. "That I may see!" (Mark 10:51). And Jesus answered and said what will you that I should do unto you? The blind man said unto him, Lord, that I might receive my sight. And Jesus granted it and he could see again.

Not only does the spirit seek life it seeks to notify of a cure in life to ills. I can say from experience the realities of these spiritual matters are real and to know where you stand is important. How do we know where we stand with God? Where do you stand in the belief of the words of Jesus Christ? In the New Testament If you

---

5. https://www.tfp.org/the-grace-of-god-still-operates-in-a-world-of-postmodern-chaos/

believe he is the son of God that's a good start It means you're willing to find out. Consider yourself very fortunate to get even to this point. There are many cannot approach the question let alone answer the question. There are also many who think they know the answer before asking the question. A perfect example of this is the cross Jesus was hung on. Up and until that day Jesus made known to the leaders and the people by good deeds, he was the son of God sent to make straight the crooked *Paths*. This is a huge change and moment in time because it's when God accepted the sacrifice for SINS created by the war in heaven and the devil cast out unto the earth first in the Garden with God's spiritual creation man and women then cast out of the garden also for beguiling his creation which could have removed the sin right there with the tree of life present and able to accept repentance from the devil. Who is the tree of life? Jesus Christ the personage of God's word. Jesus said I am before the foundation of this world. The Jews were blinded that we the gentile or any other person on earth with a soul, if they believe in Jesus Christ shall be saved, they that believe not shall be dammed. It's a yes or no question. We the saints of the earth fight not against flesh and blood but spiritual battles for others and ourselves as we are tempted daily by voices and circumstance to cause us to stumble. Prayer for others, the leaders to uphold the standard against evil it's a command from God himself Ezekiel 33:5-11 is a good example of this. Warn the people or their blood be upon you and God takes no pleasure in the death of the wicked only they change to meet good and find everlasting life. What a message is in that we have power from God through our prayers and testimony and witness that establish that same word of God from the beginning. The word needs no reestablishing its always and will always be among those that follow that word for the word is God. Jesus said, "Heaven and earth will pass away but my words will never pass away." What man expects is God to belay sentencing, put off the judgment, the indictment of all who have purposefully denied good, denied truth, and denied everlasting life. The indictment which is coming is not for the saved it's for the dead while they live, for the blind that want blindness, for the covering of the truth and think they do God a service. When clearly the precepts through time say otherwise that good is to be followed with the neighbor, with the family, with the spirit of life. If you are taught to observe the daily love of God, and someone comes in says try this for a change and its bad and out of line with your teaching, do you, do it?

How many souls sided with the devil when he was cast out? We have a clue ⅓ of the heavenly host was cast out with the Devil and were outside that garden in the beginning because the garden had four personages. In the course of millennia on the course to understanding the size of the group with the devil we find Moses born to establish the laws of God. One interesting point is archeological evidence of some 2 million left Egypt but only two made it to the promised land of the original group, Joshua and Caleb. When we understand the Mesopotamian and Phoenician epochs, we find warring people always set against God so how big are these groups? Today we have 7.8 billion souls or there about. When we use the 2:2 million ratio we find the saved souls that will side with truth. From those there is the lukewarm as Revelations 3:14 that end day church Laodicea or the people Lao deceived dicea this is the last age. Who will know truth and declare it? And that age will be naked blind and don't know they are naked or blind, terrible condition. Need of nothing, heady high minded, doers of evil rather than good and think they do God service by their deeds of evil to proclaim there is no God and squarely sets the heritage of their own hand to build the final separation the abomination of desolation using *man to kill man*. I say this, stand on the side of the Lord Jesus Christ and find peace to practice Love, Peace, Truth, Charity, Compassion, which bring Strength, Power from the author of life.

God is the God of the living not the dead. Past churches present churches, past leaders, present leaders, those who judged others that beheaded others will be indicted by the spirit of testimony to this age. Then will those two witnesses Daniel and John stand up declare the power of God. At that time, you will see visions, un natural phenomena and even the angels themselves will appear. What are they? They are servants of God to bring in the day of judgment. The only way out of that judgment is by Jesus Christ who is able to remove the curse, remove the blindness, remove the pains of death and usher in the truth. One thing in the logical realm is Jesus Christ has a more perfect knowledge, a standing reality to his pathway which ends in life not death.

Now today I hear the clinking of arms and sword to begin *the man killing man* again and again which solve nothing. Is man become so ignorant that his heel is bruised, and the head is bruised? So out of that garden the curse was instilled upon that serpent the devil and so it is today enmity abounds. But the devil is subtle very cunning to get what he desires the souls of God's creation by the creation itself this is the reality. Why wars? Why no peace? Why was Jesus killed for doing good? This is what must add up. The Devil in the details creates first the separation by using words which cause a direction to be observed. When its observed and found to be a force which allows people to gain control of the overall narrative, we then have the building blocks of hate mistrust and all the evil pathways with reward. A temporary easy life and money lots of blood money. The depth of the delusion is evident today as wars are being fought for. And money is being used for? Ask this question, how much money goes toward the defense of the USA? $750 billion nope its way higher those are just superficial numbers. So, the Devil is doing all he can to edify evil and evil responds with deeper separation deeper divides. Look at masons, who are these who challenge God? They are the servants of the Devil, and their degrees are literal agencies against the good to overcome and eliminate them, but so doing they lose their souls. They don't lose their spirits however, because the spirit is the testifier and witness in truth. So, these that ply the trade of evil know what the cost is but are foolish to sell that truth. These are they that need prayer as they are leaders of countries, kingdoms, principalities.

What is the greatest understanding on this planet? It's knowing who we are and why we are. Being tested is perpetual generation to generation and soul to soul until the final days are consumed. Reincarnation of a soul is for the purpose of perfecting what God sees as the gift. Who are the so-called aliens? These are civilizations which found life as a peaceful meaning to learn the universe and its workings. These are the reality port of our existence and there are many eyes on us to see if we destroy all humanity or transcend to life and peace. This is the true war which rages in dimensions of space and time. Jesus is the head and giver of life the personage of the word of God that all who believe on him shall be saved from the second death the everlasting separation from life.

I've listened to many JFK speeches and Kennedy indeed said these very words. So, if people claim it never happened I have the voice recording of his saying so about slavery. The murder of a man for his belief in people and courage to say it is in my book the highest honor a man can give of himself. The cowards who still today practice their agenda to enslave is very real and only a few steps from becoming law of the land. They will take your guns, take your dignity, take your truth and laugh at you while they do it. You know who they are, and they have absolutely no shame from you and causing wars to take your freedom, liberty and rights through separation dividing the lot until the poor are the slaves, no middle class. Only the rich or what they think is rich will be faced with a second death and the books will be opened and judgment commence. May the Lord have mercy to them on that day because it's the last separation they will ever see.

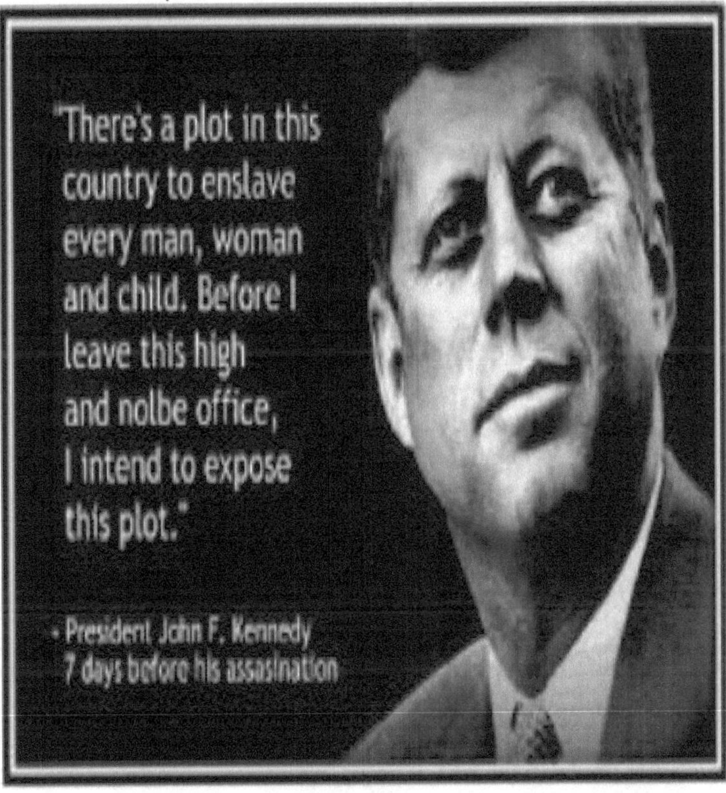

So, I ask, are we safe? Are we watching the same deliberations? If you are the least bit curious and want to join our freedom to be free, would it not be in our best interests to fly with birds of the same feather? If we allow the tyrant to take our little freedoms left, we will be face to face in a fight against each other because that's their end game. We simply have to ask ourselves do we want tyranny? Or do we want good pathways which promote equality for real.. ! We are many more than they who regard us as servants, and slaves. Servants are paid but slaves are used. Planetaryhealthproject.org is a starting point for realistic future good. I hope to see many of you there and we work with law groups and freedom groups which send the message we are the world, and we are free.

It is with great respect I write this book for the people to see truth.

Did the leaders offer a seat for the reality check and change for the better? COP26 we were completely ignored, but why? Because what we are bringing is the solution not fabrication. The whole world concluded these COP meetings are totally useless in time and resources to get them all there. Imagine John unfolding the reality of his water invention.. ! This invention capable of powering the entire planet in a short period of time to meet future need. Now we have Russia causing oil supply to be disrupted because of war. Are you beginning to understand these leaders are delusional.. !

Are you beginning to see the repetition of failed pathways and look these people have lots of money that makes them dangerous? But they will press the poor and give less and less material to the poor to keep them dumb and under the slave theme. One should be outraged by now to the whims of the rich. Stand up and learn and back the petition in this book it's time to do the right pathway. As they say the proof is in the pudding and at every turn, we become less able to function our rights. Join Richard in his quest for Time-Equity here https://www.google.com/url?sa=t&rct=j&q=&esrc=s&source=web&cd=&cad=rja&uact=8&ved=

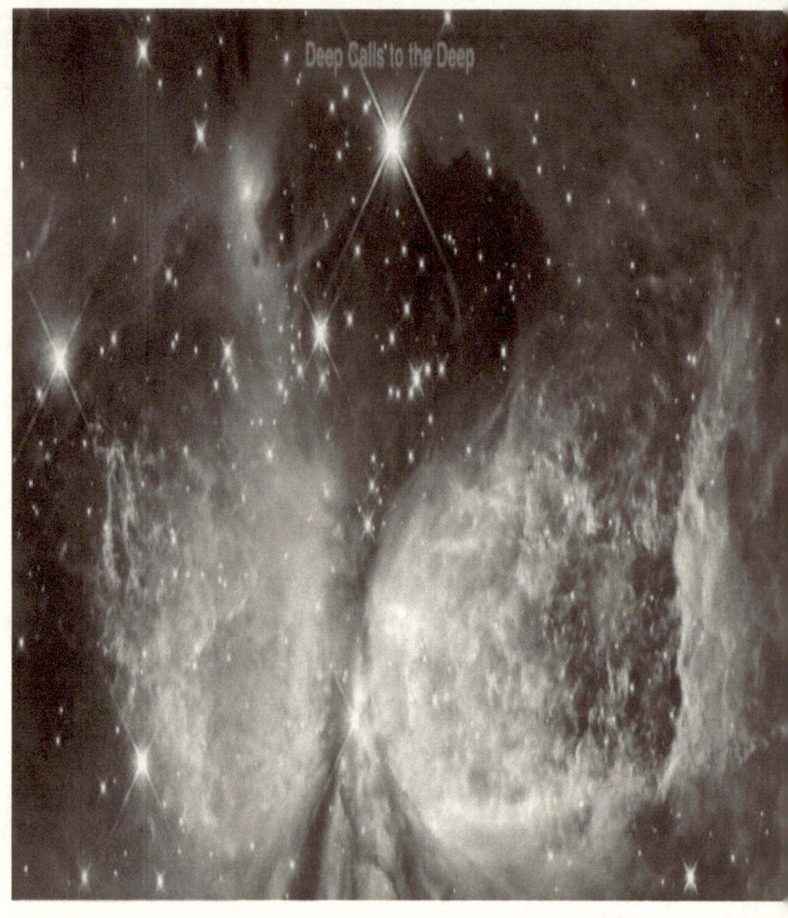
Deep Calls to the Deep

Are we ready to find God and find him loving us as he said he would? I believe there is a great transformation going to take place. All the new sightings taking place not only on earth but on the moon what would be described as three ships 5-15 miles long and 3 miles wide.. !! now that's a huge story of what is going to take place. These ships are the next level for those that love peace, truth, charity, compassion which offer strength and power. We are at the door fellow world citizens and the oligarch, the evil, the haters are all going to be round up and the new era begins right after their monolithic attempt to rule the world. Look to 2027 August and I'm with that new era management Jesus Christ who prepared a place for us. How right and fitting is that?

References

⌘

https://uaustin.org is a bold new way of education
https://www.stopworldcontrol.com/planned
John Rosebush Information
https://www.youtube.com/watch?v=wa2WZcmpzYw
https://www.youtube.com/watch?v=1PMG4ZcNR7s&t=3s
https://www.youtube.com/watch?v=BUbyuKjtHss&t=295s
https://johnrosebush.blogspot.com/2018/03/worldwide-development-corporation.html
World Economic Forum world take over
Agriculture serious issues Part 1

Part 2

Pfizer known results but carried on anyway
  2010 Paper on Pfizer knowledge points
  Richard Keirnicki
  richard@richardmkiernicki.com
  <u>Ncareersobel Prize winning Drug</u>
  <u>Public Health and Medical Professionals for Transparency</u>
<u>Public Health and Medical Professionals for Transparency</u>[1]
  <u>https://www.medalerts.org</u> To know the numbers National
Childhood Vaccine Injury Act of 1986
  <u>They tried to assassinate him</u>[2] that tried to assassinate him
  <u>Hale leaked documents</u>[3] Hale leaked documents
  <u>https://planetaryhealthproject.org</u>

---

1.   https://phmpt.org/

2.   https://thepulse.one/2021/09/27/the-cia-wanted-to-assassinate-julian-assange-a-new-investigation-reveals/

3.   https://thepulse.one/2021/08/02/whistleblower-jailed-for-showing-90-of-people-killed-by-u-s-drones-are-bystanders/

Epilogue

What countless societies and peoples and nations and kingdoms poets, philosophers have sought to find in lasting truth which binds a civilization is evident in these writings. We continue to grope for the reality that can bring substantial trust for a brave new world seeing afar off, but never able to bring together fully. The forces which bind nature cannot bind the mind of evil therefore its left up to the good to converse among good to overcome that evil which stagnates and pauses the forward truthful pathway.

We can no longer afford the mistakes of evil intent, nor can we welcome the hate the evil brings. There is a deep out there which defines our good pathway, good intentions and good actions one to the other. It is my intention to succeed with one goal in mind that defines that reality existence to push the boundary of life to respected existences in all cases. These forces now threaten to engulf an entire civilization which was given the best chances to overcome illogical structure with logical base narratives of the philosophic and angelic. The many cases from Jesus Christ to miraculous healings of William Branham which pointed always to salvation, and everlasting life.

The world rejects such as these information portals but now there are the pictures of the downtrodden there are the evidence of a deeper course flowing around us which is free to all to receive when they accept the reality there is more than what meets the eye. I merely present the proof, but more than proof a challenge which will save the reader from the clutches of Evil intent and beguiling of your senses when presented with riches and need of nothing. This book testifies as a witness to the indictment which will be handed down from God to his two witnesses Daniel and John. As is said, they that believe shall be saved. I need not mention the rest for it delivers the promise either way. Please believe be saved and do good and support life which sets us as sons and daughters of God to satisfy the transformation from death to life for God is the God of the living. The next examples are two of thousands of healings this man did in the name of Jesus Christ. Are they all coincidence? Why are they not known?

Girl 'Cured' Second Time by Healer, Only 'Miracle' Here

Lame, Halt, Blind Leave Meeting As They Arrived—Full of Hope

By CHARLES MacFARLANE

They came on crutches, in wheelchairs and on stretchers.

They left on crutches, in wheelchairs and on stretchers.

But still they had faith, faith in Rev. William Branham.

WRITER'S SON BOOK-THIEF?

HOLLYWOOD, Nov. 6—(UPI)—Paul Rivas, 18, son of Humberto Rivas, Mexico City film writer, was held today on suspicion of burglary for allegedly taking books from hotel rooms he

Boy Raised from the dead and girl cured of Leukemia.
These stories of the past will again take place as God said he would
pour out his spirit upon all flesh in those last days. Acts 2:21 And it
shall come to pass, that whosoever shall call upon the name of the
Lord shall be saved. This is not maybe talking. Are you ready to see
Jesus Christ, let us not be sleeping, let us not be afraid, let us find
the assured promise they that seek shall find? Remember this past
brother passed all the tests bringing miracles before the people and

hardly anyone knows about the man and gifts. I present the evidence go see for yourself these evidence.

In reality it takes a fortified spiritual understanding to weight the gravity of the situation with respect to evil and evils intent. I will explain this here for the benefit of the reader. Evil and the Devils under satanic forces being generated by hate for the purpose of man to destroy man. Satan never wanted the creation of God to go further than that original garden. All Satan's efforts go toward the collapse of moral integrity and foundational thinking which forms good pathways. This is the primary reason we are delusional with money and its importance. Think a moment we now have added 660 billionaires during a pandemic crisis created by the elite that love money. We now have some 2755 of them.. ! using 1.7 earths the days are numbered.

Jesus said the root of all evil is the love of money. A person that loves money will do anything for it and anything to keep it. If you feel this is all just a ploy, I assure you the final days are completely ready to be accomplished. Some will wake their fire inside to pray for the lost, for the leaders being coerced and separated. If you do this, it will be well with you and God will recognize you as a son or daughter and the fiery darts of Satan can no longer touch you. Be vigilant to recognize the reality I have presented with the many proofs and actions written here. There is Love, Peace, Truth, Charity, Compassion which uphold Strength and Power not the other way around.

Don't miss out!

Visit the website below and you can sign up to receive emails whenever James Boyd Fuller publishes a new book. There's no charge and no obligation.

https://books2read.com/r/B-A-MCRZ-ZPCMC

BOOKS 2 READ

Connecting independent readers to independent writers.

About the Author

James is a mechanical engineer for 22+ years who understands power systems and what it takes to make real world power generation of up to 1gw in the space of a football field. The design is ready to be deployed for cities or wherever nedded. 1/5th the cost of solar or wind this technology will emmensly help our civilization move forward out of the failing current pathway.

Read more at https://jamesboydfuller.com.

www.ingramcontent.com/pod-product-compliance
Lightning Source LLC
Chambersburg PA
CBHW031247160726
47993CB00001B/64